Field Guide to
WEEDS

Based on
Wildly Successful Plants
by Lawrence J. Crockett

With illustrations by
Joanne Bradley

A MAIN STREET BOOK

This field guide is not intended as an authoritative source of information on the edibility
of weeds and wild plants. Those wanting this knowledge should seek more information
by joining a local club or society, reading many field guides, and consulting with local experts
on the identification of any species found. Neither the publisher, editor, nor the reviser
are in any way responsible for mistaken identifications and /or the consumption of any toxic
or deadly species that may occur to anyone who fails to read or heed this warning.

Text for this book was adapted and updated from
Wildly Successful Plants
© 1977 by Lawrence J. Crockett
Line illustrations by Joanne Bradley

Photography by
Larry Allain, Jennifer Anderson, Steven Daniel, William S. Justice,
Dr. John Meade, Robert H. Mohlenbrock, John M. Randall, James L. Reveal,
George F. Russell, Carol Southby, David T. Southby, Bill Summers, and W.L. Wagner

Editor: Cassia B. Farkas
Design and production: Alan Barnett, Inc.

10 9 8 7 6 5 4 3 2 1

© 2003 by Lawrence J. Crockett
Published by Sterling Publishing Co., Inc.
387 Park Avenue South, New York, NY 10016
Distributed in Canada by Sterling Publishing
c/o Canadian Manda Group, One Atlantic Avenue, Suite 105
Toronto, Ontario, Canada M6K 3E7
Distributed in Great Britain by Chrysalis Books
64 Brewery Road, London N7 9NT, England
Distributed in Australia by Capricorn Link (Australia) Pty. Ltd.
P.O. Box 704, Windsor, NSW 2756, Australia

Sterling ISBN 1-4027-0694-4

CONTENTS

Preface

by Jaqueline B. Glasthal

Creating a system of classifying weeds is truly a tricky business—particularly since there is no consistent definition of what constitutes a weed. Any plant that grows where it isn't wanted falls into this category. But what is unwanted—and where it is unwanted—varies a great deal, regionally, seasonally, and even at different times in history. By its basic definition, corn that crops up in a soybean field is a weed, as much as are curly leaf pondweeds that grow out of control in the eastern and mid-western United States. In Washington State, however, curly leaf pondweeds have not caused problems and thus do not have the same "weed" status as they have been given further east. As Lawrence J. Crockett noted in the original Preface to this field guide, "It is very likely that our basic agricultural plants began as invading weeds thousands of years ago. Slowly (and probably accidentally) the primitive early farmers observed that the invader offered a far better food source, [making] greater civilization possible."

Today, a large portion of those plants classified as weeds in North America are non-natives, meaning that they originally came from elsewhere. The rest just have a knack for rearing their pretty heads where landowners and others wish that they would not. Though some are quite beautiful, these "wildly successful plants" can still be the bane of the homeowner or garden grower's existence. Yet, since it seems that weeds are all around us and are here to stay, it makes sense to get to know them better. In the process you may even find that, when kept in check, they can be quite useful. Dandelion greens, for example, taste great in a salad; their roots, when dried and ground, can serve as a substitute for coffee; and their blossoms, when steeped in water, make "dandelion wine."

For the novice not trained in botany and taxonomy or the ways and habits of plant classification, however, identifying various species can be difficult and sometimes frustrating. There is a special jargon for various taxonomical systems, and many of these terms have several shades of meaning, not immediately apparent to those not in the know. For example, when reading that a twig is "stout," you may be left wondering: just how stout *is* "stout" anyway?

Given his appreciation for this dilemma, Lawrence J. Crockett made up his own classification "system," which has been retained here. As a professor of biology at the City College of New York, a long-time member of the editorial board of the American Journal of Botany, and a past president of the Torrey Botanical Club (now the Torrey Botanical Society)—which honors the contributions of the great nineteenth century botanist and plant taxonomist John Torrey—Crockett was perhaps the perfect candidate to compile this field guide, devised specifically with the amateur botanist and the plants in this book in mind. Though it was unrealistic to omit all specialized terms, the amount to which they are relied upon has intentionally been kept to the minimum. And when found to be unnecessary, these terms have scrupulously been "weeded out."

The best way to use this book, then, is to start with the introduction that starts on page 7. It will give you some crucial background information about plants in general, as well as the particular traits and characteristics that you will want to be on the lookout for. Since a picture is often worth a thousand words, the illustrations and photographs used throughout should also prove useful as will the glossary provided at the back of the book. Even if you are not currently trying to figure what that is growing in the middle of your tomato patch or otherwise perfect lawn, this volume has much to offer. As the original author did, we think you'll enjoy simply reading about these plants for their own sakes—and for yours.

Introduction

ALL YOU NEED TO KNOW ABOUT WEEDS

There are no weeds in nature, just as there are no peasants. Civilization and cultivation have created both. It is obvious that botanists have had some difficulty defining the concept "weed." In general, they have come to the conclusion that a weed is any plant found growing rapidly and abundantly in a place where it is not wanted. Perhaps when the ancient farmers cleared a small piece of land, planted a crop, and then had to fight some invading plant species, that invader became the first weed. Or was the first weed a plant that, because of its rapid rank growth, contested with men the entrance to their first cave homes?

There are no records to tell us when human beings first confronted what they came to call weeds. Records of weedy plants, even from peoples who kept good records (such as the ancient Egyptians), are scanty and apparently no early civilization had a clear-cut word for a weed.

Our word "weed" descends from Middle English (by way of, and influenced by, Flemish, Anglo-Saxon, and Frisian), but the best etymologists think that the origin of the word has been lost to us. Some believe it is related to the Old English word "woad," applied to the leguminous plant that used to supply a valuable blue dye to the English, but this is by no means a clear fact.

If a weed is defined only by its unwantedness, much is being overlooked by the definer. Let us consider the most hated weed plant of the suburban lawn grower: crabgrass.

Digitaria sanguinalis (p. 66) is the botanist's name for crabgrass, but in English or Latin this plant is the enemy of the householder and the mere mention of its name will bring blood to the whites of the eyes of suburban gentlemen. Yet the strawberry grower puts the same grass to good use (p. 66), and so does the Southern farmer, who cultivates crabgrass for forage and pasturage. So one man's plant enemy is another's good friend. There are many, many weeds in this category.

Almost all the definitions of "weed" depend on the opinion of the one who is doing the defining. Certain characteristics of weedy species are more personally important to the definer than to others, and therefore definitions that include a number of different attributes of weedy species have arisen. These have included: their danger to man or animals; their rank, rapid, wild growth; their successful growth, unbidden, unsown, and uncultivated; their usually aggressive and competitive growth habits; their insistence on growing where they are not wanted; their persistence and resistance to control or eradication; their ability to reproduce themselves rapidly both by seeds and by vegetable means; their penchant for making the landscape less attractive to the human eye. From this list it is easy to see that man is very personally involved in his definition of the concept "weed" and that a number of his reasons for disliking weedy species are psychologically colored.

Not all those who have written on the subject have dealt with weeds as unwanted or "evil" plants. A few voices have been raised in their favor, most prompted by the ecological point of view, i.e., the disposition to take the plants as plants.

A. H. Bunting said in 1960, "Weeds are pioneers of secondary succession of which the weedy arable field is the special case." F. C. King wrote, "Weeds have always been condemned without a fair trial," and Joseph Cocannouer, who has written popularly and favorably about weeds, states, "This thing of considering all weeds as bad is nonsensical!"

There is considerable evidence that many of our major agricultural species, e.g., wheat, maize (corn), rye, and rice, entered the field of agriculture as weeds, growing alongside other plants, at the time considered more useful. Lamb's-quarters [*Chenopodium album*, p. 86], one of the most common weedy flowering plant species, was apparently a plant grown by our ancestors for food. Both its leaves and seeds are edible. When lamb's-quarters was an important food plant, the ancestors of wheat, maize, or rice (long since become our major agricultural crop plants) may, indeed, have been growing along the borders of cultivated Chenopodium patches, or invaded those patches and had to be removed as "weeds." Perhaps by accident (a lazy early farmer?) it was discovered that the seeds of grasses were even more valuable than those of lamb's-quarters.

The gardener or preserver of the lawn or any other nontechnically trained plant lover cannot be expected to become versed in ecology or ethnobotany before trying to understand the nature of weeds, but he should be aware of some interesting facts about weeds.

By and large humanity has itself to blame for the weedy species that follow in its steps. Have you ever noticed that the most favored habitats of weeds are open, sunny places? Few weeds are tolerant of shade. It was man who opened up the sunny spots where weeds are happiest. Certainly there were open spots when the forests of the United States were intact, but these had been open for ages and the plants in them had attained a definite ecological balance. Few open areas in the United States have this balance today. Man tore down the forests, cut swathes through grassland, forest, and other long-established ecological situations, and cut open pathways for roads and railroads across the entire nation. His clothes, vehicles, and animals all carried seeds and liberally scattered them as he himself spread across the continent.

This is not to suggest that only man has opened areas suddenly, cleared large patches of the forest or grassland of all their established vegetation. Glaciation, fires, floods, and other natural disturbances have occurred in the past, and will occur in the future, but man is the greatest and most persistent disturber of established plant growth patterns. Gardens are not natural; farms are not natural; lawns are not natural. No, man is the chief disrupter of long-established ecological plant relationships.

Man and the weedy plants are closer friends than he generally seems to know; indeed, his civilization may depend on that friendship. The human being and the weed are wed, though to listen to most people talk about weeds, one would think the relationship was out of *Who's Afraid of Virginia Woolf?*

It is interesting to note how few weeds are native to the United States. When the white man began to colonize this country, he brought with him familiar vegetable plants, kitchen herbs to spice his foods, medicinal plants to protect him in the frightening New World, and garden ornamentals to remind him of the old sod and its floral beauties. After he was here a while, he found that his adopted land had plants that could serve all these purposes and he included these in his farm and garden. The plants he had brought with him made their escape over fence or wall and, as species happiest in wide open spaces, became naturalized weeds, such as *Linaria vulgaris* (butter-and-eggs, p. 106).

Other species came unbidden and unsuspected. They came as seeds attached to the clothes of arriving colonists or to the fur of imported farm or domestic animals. Soil used as ballast in English ships in the eighteenth century was dropped in our eastern port towns when rich cargoes were taken up in its place for the return trip to the mother country. From that ballast hundreds of new plant species were introduced to this country.

The noted weed expert J. M. Fogg has found that 14 percent of the plants listed in Gray's *Manual of Botany* are introduced plants. Somewhat more than half the foreign species came from Europe, and Eurasia and eastern Asia together contributed about a third as many as Europe. Tropical America has donated less than one hundred species.

Certain families of flowering plants predominate in the numbers of weedy species they contain. They are, in order of the numbers of weeds they have produced, the composite, grass, mustard, legume, pink, mint, and snapdragon families. The overall majority of nonwoody plants, and all the families listed above, are relatively recently evolved: the composite, mint, and snapdragon families are thought to be among the most recently evolved plants on earth. Weeds deserve some credit for their pioneering ability; they certainly suggest genetic hardiness.

In fact, the aggressiveness of introduced weeds is a threat. It is their very vigor that drives out not only our garden plants but also many of our native wild species. Thus we may not define a weed exclusively as a plant we do not want where it grows in garden, lawn, or landscaped highway. Both native and introduced weedy species conform to this definition, but introduced species may, in addition, drive out our native wild species, denying them their former habitats and ecological niches. The Japanese honeysuckle [*Lonicera japonica*, p. 34] will usurp territory at an amazingly rapid rate and in all directions as soon as it has set itself up in a new location. It will then proceed to crowd out most native plants, thus reducing the variety of plant life in the area. The purple loosestrife [*Lythrum salicaria*, p. 204] is guilty of the same process in wetter areas. Both honeysuckle and purple loosestrife are lovely to look at when cultivated or controlled, but uncontrolled, they endanger our native wild plants and become weeds.

Weed Dispersal

Surely when faced with a yardful of rapidly growing weeds you have thought, what have weeds got that cultivated plants seem to lack? This question is partly answerable and partly a mystery to plant scientists. Weed species generally grow very rapidly after germination from seed, and this gives them an immediate advantage. Their roots get to the moisture-laden soil before those of the cultivated species, and their fast-growing stems and leaves deprive of sunlight those plants that are growing less rapidly around them. Garden ornamentals are often plants that are happiest in bright sunlight, i.e., intolerant of shade. Once the weed species has raised its leaves about the young cultivated plant, the latter's time is almost up; at the very least, its growth will be severely inhibited. Why roots and stems of weeds grow faster, why their roots put out more root hairs, why their germination from seed is more rapid, is fundamentally impossible to explain except in very general terms such as "genetic vigor."

On a particularly dry, hot August afternoon when the hydrangea leaves are hanging down like the ears of a beagle and the cultivated species can almost be heard to pant, take a good look at weedy plants. Chances are they won't look nearly as heat-worn as your more delicate species. The reason is that the weedy species have grown a greater number of root hairs and pushed their roots deeper and faster into the soil than have the cultivated plants. Their vigor is obvious.

Many, many weedy species produce huge numbers of seeds. This is the case with members of the large tribe of composites. Composites, e.g., the daisy do not produce a single flower, though the eye sees only a single flower. Rather, they have a head of many flowers all packed together, giving the appearance of a single flower. There are two types of flower in the daisy head: a ray flower, produced at the edge of the head, and a disc flower, yellow in color, of which the center yellow disc of the daisy is composed. Since the flowers of the head mature in a spiral from the outside toward the center of the head, seeds set over a long period of time. Other types of plant usually present but one or a few flowers at a time to the pollinating insect. Their seeds are shed at once on the maturation or opening of the fruit, or shortly after the fall of the fruit.

Some weeds are provided with special dispersal mechanisms; one such is the dandelion. Each seed is provided with its private parachute (a modified calyx) to carry it through the air. Other weed species produce fruits that are barbed or covered with teeth that stick to the fur of passing animals (or to the clothes of passing human beings) and are then carried great distances before being brushed away or scratched off.

The seeds of weedy species tend to be long-lived. In 1879 an American botanist, Dr. W. J. Beal, at the Michigan Agricultural Experimentation Station in East Lansing, tested the longevity of weed seeds in an experiment that is still running. He buried the seeds of twenty common weedy species in containers. At intervals of five years until 1920, and at ten years after 1920, germination tests were run on the weeds. Thirty years after the seeds were buried, about 50 percent of the species tested still had viable seeds. In 1960, some eighty years after the seeds were buried, seeds of three weeds were still alive, and germinated. They were weeds described later in the book: curly dock [*Rumex crispus*, p. 92], evening primrose [*Oenothera biennis*, p. 108], and moth mullein [*Verbascum blattaria*, p. 116]. And, as expected, seeds of the three were still alive when the centennial of the experiment was celebrated in 1979.

Not only do weeds produce huge numbers of seeds that are easily dispersed, but many of them also produce underground stems or rhizomes (sometimes called rootstocks). Rhizomes are the bane of the gardener. Indeed, perennial weeds (those that live from year to year) produce both aerial and underground stems and can spread even if they never get the chance to produce a single seed. Weeds with rhizomes are the most difficult to eradicate. The aerial stems may be cut away or chemically poisoned, but if the rootstock is left undamaged, the plant will continue to produce aerial shoots.

The rootstock of some weeds seems to delight in being fragmented, each piece giving rise to a new, complete plant. Cutting back the aerial stems, which make the foods stored in the rhizome, will eventually starve the underground parts, but this is a tedious, time-consuming task. Chemical attack on the rootstock is not always easy or successful either.

The aerial stem may also be involved in the direct spread of a weedy species. Aerial stems of many species take root wherever they touch the soil, especially at their nodes (where leaves appear on the stem). Essentially, a new plant is formed, which itself then sends out runners. If such a plant also produces abundant seed, it is a difficult one to beat!

Annual weeds live but one year and then set seed. Death follows soon after. These are the easiest weeds to defeat if you cut the stem before seed set. *Winter annuals* germinate in the fall, producing a root and a few leaves. Then, having a head start the next spring, they produce their flowers and seeds early.

Biennials live for two years, usually producing a rosette of leaves during the first year. The rosette is produced on the surface of the soil. A flowering stalk arises during the second year of growth. If these plants are torn out or killed with chemicals during their first year, they will no longer be a gardening headache. But if you let them set seed, though they themselves die at the end of the second season of growth, your problems will have multiplied many times and will lie completely hidden until germination time the next year and for many years to come.

In the pages to follow there will be much information about weeds and their ways.

Some Basic Facts about Flowering Plants

All but two of the weedy species presented in this book are flowering plants or *Angiospermae*. Therefore the contents of this section may help you to understand the nature and structure of *all* flowering plants, weedy as well as nonweedy species.

Flowering plants usually possess an aboveground *shoot system* composed of stem, leaf, axillary or lateral buds, eventually a flower or flowers, and, at the very tip of the stem, a shoot meristem or growing point. Below ground level is a *root system* of which every branch is tipped by a root meristem or growing point.

If the plant is *perennial,* i.e., grows from year to year for more than two years, it will have within its stem and root a means for expanding laterally or in girth. Here will be found special growth tissues called *cambia.* There are two such cambia within older roots and stems. One is the *vascular cambium,* which produces new conducting tissues, *xylem,* by taking water and minerals through the root and stem (usually in an upward direction), and *phloem* (pronounced as if spelled *flome*), by carrying manufactured carbohydrates and other substances from the leaves where they are made to the other parts of the plant (usually in a downward direction). The other is the *cork cambium,* which gives rise to new storage tissues in stem and root (called cortex), and to the most external tissue (the cork), which protects the outside of older roots and stems.

Thus when you point to a tree and say, "Oh, look at that interesting bark!" you are not entirely correct botanically, for what you are indicating is not the bark but the cork. Scientists must define everything very carefully, and to plant scientists or botanists, bark includes all tissues from the internal tissue, the vascular cambium, out to the cork, or external tissues.

The root system, too, has apical meristems or growing points on the tips of all its roots. Cells, constantly formed within the meristems, later mature as various root tissues (xylem and phloem and cortex are examples),

and the increase in cells pushes the root down through the soil, while the shoot meristems keep the shoot pushing its way through the air. Hormone systems keep each organ growing in the correct (and favorable) direction. Thus the vertical growth of the plant is accomplished by its meristems (root and shoot). Growth accomplished by the meristems at the tips of the root is known as *primary growth.* Since plants called annuals live for only one growing season and do not develop cambia, they are produced entirely by primary growth. So, too, are biennials, which live for only two growing seasons. The pigweed [*Chenopodium album,* p. 86] is an annual, and mullein [*Verbascum thapsus,* p. 114] is a biennial.

While annuals are composed fundamentally of tissues produced during primary growth, perennials include an additional kind of growth. Not only do they grow by means of their shoot and root meristems, but they also develop cambia in both systems and thus grow laterally as well. Growth produced by cambia is known as secondary growth. Perennials have functioning cambia by the end of their first year of growth. The tree-that-grows-in-Brooklyn [*Ailanthus altissima,* p. 28] is a perennial.

Within the perennial plant it is the annual increase of xylem tissue (which forms as a ring of tissue) that permits botanists to determine the age of the tree by counting the rings. Annual rings also give clues to the ecological conditions through which the tree has passed during its years of growth.

Anthropologists now use tree rings as part of their technique for determining the age of villages in the American Southwest, method first worked out by Professor A. E. Douglass of the University of Arizona.

One more word about growing points of the stems. This region has been called a region of *continued embryology* and it proves plant growth and development are clearly different from what is found among animals. In a five-thousand-year-old bristlecone pine [*Pinus aristata*—definitely not a weed] the cells of the growing points of root and shoot are the direct descendants of those cells alive five thousand years ago at the tips of the oldest branches and roots of this botanical Methuselah.

The stem of the perennial plant remains all winter and does not die back to the ground. While its leaves usually fall away with the approach of autumn, left behind is the whole woody branch and its buds, both axillary and terminal (if a terminal bud is present); leaf scars, where last summer's leaves were attached; and cork on the outside, protecting and preventing water from escaping and bacteria from entering. The cork provides another help to the living tissues within the stem (cork cells are, by the way, dead). Small regions of looser cork cells, appearing as tiny clearer areas of varying shapes—the *lenticels*—permit the exchange of gases so necessary to maintaining life. Through the lenticels pass carbon dioxide and oxygen; the former leaves and the latter enters. Both are involved in respiration. Carbon dioxide is a waste product of this energy-storing process and oxygen is consumed. All living cells, in the plant stem or anywhere (in the human body, for example), must carry on respiration or

perish. Plants, unlike animals, also carry on photosynthesis. Respiration is a continuous action; photosynthesis is carried on only in the light.

The twig may be capped by a *terminal bud*, or the last axillary bud formed during the previous growing season will take over and act as a terminal bud, as in the tree-that-grows-in-Brooklyn [*Ailanthus altissima*, p. 28]. This kind of bud is called a *pseudoterminal bud*.

When, in the spring, new growth begins once more, the bud scales around the terminal bud will fall and the contents of the bud will expand. A line is left about the stem marking where the bud scales were attached. Bud scales are modified leaves. Counting successive lines of bud scale scars is one way of determining the age of a branch, just as the more internal annual rings of secondary xylem can be used to determine the age of a tree.

Some Modifications of the Stem

Not all stems grow erect. The *rhizome* is an underground stem that sends up leaves from its "dorsal" surface and roots from its "ventral" surface. In temperate climes ferns are characterized by rhizomes, though in more tropical climates some ferns grow as erect, tall trees. Many flowering plants also produce rhizomes, and they are quite common among the grasses. Fragmentation of the rhizome produces a number of individual plants and is thus a form of asexual reproduction.

A thickened portion of a rhizome is a *tuber*. The tuber you know best is the white potato (not originally from Ireland but from Peru), which is a stem rather than a root.

Look at a good-sized potato and note the cork or dark "skin" on the outside; the "eyes," which are actually axillary buds (and can, therefore, produce a new plant if roots can be induced to grow); and, possibly, a thin stalk protruding from one end of the potato, which represents all that is left of the normal part of the underground stem not filled with stored food, as is the tuber.

Some underground stems are nubbinlike structures that are called *corms* or solid bulbs. Crocus is grown from a corm. The true *bulb*, however, while it contains a short, thickened stem, is fundamentally composed of overlapping leaves filled with stored food. Take a whole onion and cut through it longitudinally, i.e., from the leafy part on top to the flat root-bearing part below, and look inside. Here the food-rich swollen leaves are obvious, and the small, flat stem to which they are attached (and to which the roots are attached) is also rather easy to see.

Bulbils are small editions of bulbs, but they form far above the ground in either axillary bud regions or within the inflorescence. The latter method of formation is common among onion relatives. Wild garlic [*Allium vineale*, p. 54], because of its bulbil formation, can be the bane of the lawn keeper or preserver of fine pastures. When the bulbil falls to the ground, it is ready to begin growth as soon as its roots dig into the soil.

Armaments Seen on Stems

Poetry is filled with roses that have thorns. Unfortunately for poetic accuracy, the rose does not produce a thorn! Rather the skin-stabbing structure on this beautiful flowering plant is a *prickle*.. Scientists must define everything very carefully: a true thorn is a modified stem, which the skin-stabbing structure of the rose is not. In a true thorn an axillary bud grows out, gradually hardening, until finally its apex points up. Thus the thorn is a branch that has become tough and pointed. Thorns are very hard to break off because they are so intimately connected to the main trunk or branch. The rose prickle is an outgrowth of the epidermal tissues of the stem, as are all prickles; however, I prefer that the poet continue to use "thorn" (his license) since so few lovely words rhyme with "prickle"! A third armament is found on stems. This one, which can be noted on *barberry* (p. 30), is a *spine* and actually represents a modified leaf.

General Positions of Stems

The shoot system may grow *erect* (standing upright), as most are observed to do, bend over and lean in various degrees, or lie *prostrate* on the ground. This last condition is also called *procumbent—decumbent* if the stem bows nearly to the ground. A *creeping* or *repent* stem lies on the ground but produces adventitious roots (roots that appear where normally roots do not) at its nodes. (Weeds that can do this are a very special nuisance to the gardener.) If the stem can't support itself at all, it may climb as a *vine* and support itself entirely by clinging. Some vines twine about their supports—for example, the morning glory [*Convolvulus arvense*, p. 36]; some use specialized roots to hold onto the support—for example, ivy, poison ivy [*Rhus radicans*,

p. 44]; and some produce special structures called *tendrils*—for example, the fox grape [*Vitis labrusca*, p. 40]. However, the tendrils of the garden pea arise as leaflets and those of the wild cucumber [*Echinocystis lobata*, p. 42] are whole modified leaves.

Where Leaves Are Borne on Stems

Along the stems, leaves are borne at *nodes*, and the portion of the stem between the nodes is called an *internode*. Leaves are arranged on stems oppositely if the two leaves face each other, or in a circle if there are three or more leaves at a node (see *Equisetum arvense*, field horsetail, p. 206). In the vast majority of plants the leaves are borne alternately on the stems (see *Ailanthus*, p. 28). In this case the leaves are arranged in such a way that each leaf is alone at a node and at a different level from all others. Thus the leaves form a spiral about the stem.

The Green Leaf and Its Function

Leaves are the photosynthetic organs of green plants. Photosynthesis is a biochemical process that occurs only in the presence of light and chlorophyll within the chloroplast. In the process carbon dioxide, taken from the environment through the stomates (small apertures in the surface of leaves that may be opened and closed by special mechanisms), and water molecules, absorbed by the root and transported through the xylem, are reacted upon and changed within the chloroplast so that a carbohydrate and oxygen are manufactured. The oxygen is released as a waste product; our earthly oxygen blanket, upon which all animal life depends, gets most of its oxygen from this complex process! From the manufactured carbohydrate, say, glucose, the plant is capable of making all the various nutrients we need to live—fats, sugars, amino acids, vitamins—and they are available to us from no other source. If you dine on meat, you are dining on modified grass and nothing more!

Therefore, we can say all animal life depends absolutely on the photosynthetic process. If tomorrow the process were to fail globally, you and I and all other animals that live on land or in water would be doomed. All animals are fundamentally parasitic on all photosynthetic plants. Everything humankind knows and loves—our whole civilization—depends in the last analysis on the continuation of the photosynthetic process on land and in the waters of our planet. The time has come to give more thought to this: green or photosynthetic plants are the keystone of the food chains of this planet Earth.

The Structure of the Leaf

A great number of leaves are composed of a flat, expanded portion called the *blade*, a region of attachment to the stem called a *petiole*, and in the upper angle formed by the petiole and the stem's meeting an *axillary bud*. Indeed, to be a leaf the structure must have a bud in its axil. In members of the Dicotyledonae (flowering plants are divided into two large subgroups of which the Dicotyledonae, or in its shortened form, dicot, is one) the leaf meets the stem by means of the petiole itself, usually a thin, sticklike portion of the leaf. In the Monocotyledonae (or mono-cot) petioles are usually lacking and here the leaf base wraps itself around the stem or clasps it. This condition is known as *clasping* or *sessile*. In the leaf of the dicot, leaf blade veins can be seen running this way and that. This condition is known as *netted venation* or *reticulate venation*. Within the leaf of the monocot the venation is quite different. The veins run from the tip of the leaf to the base and are parallel. The shape of the monocot leaf is usually like that of a sword, i.e., long, thin, and tapering from the base to the tip, while a typical dicot leaf is elliptical. There are, however, many exceptions and leaf shape varies tremendously among the dicots.

The manufactured foods are shipped from the cells in which they are made (chlorophyl-lous cells) through leaf veins to veins in the petiole, and then into the veins in the stem and to other parts of the plant that are not photo-synthetic. The plant usually makes excess food and this is shipped to storage cells. Much food is also stored in seeds. The manufactured foods travel in the phloem tissue.

With a few exceptions, members of the monocots do not produce cambia in their stems and roots and thus do not show sec-ondary growth. The palms are one exception. Thus, no matter how thick the stem may appear, e.g., as in the "trunk" of the banana "tree," it is composed of a relatively thin stem that is made fat with overlapping, clasping leaf bases. Tulip, lily, crocus, onion, gladiolus, and grass are common monocots.

The Axillary Bud

Within the axil of the leaf is found the lateral bud. In perennials, when the leaf falls at the approach of winter, the axillary buds usually become more easy to see than in summer, when they are hidden by the leaves. This bud usually contains a complete branch system (a shoot system) and sometimes an inflores-cence as well (or alone). The axillary bud will be activated only if the overall hormonal bal-ance of the shoot is changed or disturbed. If the terminal region, the shoot apex (or termi-nal bud), is lost or destroyed, the buds lower down the stem will become active. For thou-sands of years the gardeners of Japan have used this method to control the shaping of trees and shrubs, though only in more recent times have botanists begun to learn what really is occurring when a shoot apex is removed from a branch.

Simple Leaves and Compound Leaves

When the leaf is all of one piece, we say it is *simple* or *entire*. An entire leaf may have veins that branch out horizontally to each other from the main vein or *midrib*, in which case the entire leaf is said to be *pinnately veined*. Or the major veins may arise from one point at the base of the leaf where it meets the petiole, and then radiate like the fingers on a hand into the blade. Such a leaf is said to be *palmately veined*. (The leaf of the common household geranium has such a venation.)

The Compound Leaf

The preceding description dealt with a leaf composed of one unit—a simple, entire leaf. Not all leaves are so composed. To understand the structure of compound leaves, keep clearly in mind that to be a true leaf—by botanical definition—the leaf must have a bud in its axil.

Now, imagine a large, entire, pinnately-veined leaf. Then begin to imagine that the blade tissue around each of the horizontal veins separates from its sisters until finally each horizontal vein is surrounded by blade tissue, but entirely separate from all others like itself, though all are attached to the midrib. You have created the image of a giant feather, or a *pinnately-compound* leaf (*pinna*, as in pinnate, means "feather"). The axillary bud is found only at the base of the petiole; none will be present at the base of the newly created *leaflets*. The leaf of species of the genus *Rhus* (sumacs, p. 44) have pinnately-compound leaves.

You may imagine, using the same rules for leaflets as you used for the whole leaf, that a pinnately-compound leaf could be compounded several times; if twice, the leaf is *bipinnate*. Soon such a leaf is very plumelike.

Let's, once more, play divine. Begin this act of creation with an entire, simple, but palmately-veined leaf. Now imagine that the tissue about each fingerlike vein separates from its sister veins as in the pinnately-compound leaf, leaving the new leaf composed of fingerlike leaflets radiating from the most distal portion of the petiole. This is a *palmately-compound* leaf. The relatively well known horse chestnut tree [*Aesculus*

hippocastanum] has palmately-compound leaves. The axillary bud of these leaves is found only at the base of the petiole, never at the base of a leaflet.

Other Kinds of Leaf Differences

Even more variation occurs. The margins or edges of leaves or leaflets show a myriad of shape variations. Here are but a few: *undulate* (wavy), *serrate* (with teeth pointed forward), *dentate* (with large teeth).

The whole leaf may also vary in shape from species to species—and sometimes on the same plant! Leaves may have an overall shape that is *orbicular* (round), *ovate* (egg-shaped), *lanceolate* (lance-shaped), *reniform* (kidney-shaped or bean-shaped), *sagittate* (arrowhead-shaped), *deltoid* (delta-shaped or triangular), *spatulate* (shaped like a spoon)—and those are but a few. All of these differences may be of great use to the botanist in making scientific classifications of plants in erecting keys for their identification.

Roots

I have always felt somewhat sorry for roots. Unseen, they are often ignored. Yet no plant could do without them, except for the few specifically evolved to go rootless. Most of our weed species are provided with very healthy root systems—indeed, they apparently have more actively growing roots than most of the more desirable garden species. Among the plants there are two major types of root system produced: the *taproot system* and the *fibrous root system*. The carrot best typifies the taproot system in which there is one main, conspicuous, dominating root and numerous, sometimes almost hairlike, secondary roots. Select a fresh carrot and study it closely. Note that the bulk of the carrot is one unit, the taproot, and that small, thin side branches seem to come out from wrinkles in the main root. These are the secondary roots. The carrot of commerce [*Daucus carota*] is the same species as Queen Anne's lace (p. 172). Dandelion [*Taraxacum officinale*, p. 104] also has a taproot.

The fibrous root system is typified by that seen on grasses. Here one root does not dominate the system; rather, all the roots are equal, and there are a great many of them. Gingerly pull up some grass (preferably not from your own lawn), and you will find the fibrous root system. You will also find that in pulling up one plant, you usually pull up a region of plants because the roots of fibrous-rooted plants are interwoven. Pulling up taprooted plants may not be easy either, and you may leave a small piece way down deep that, in time, will reproduce the plant.

Other Forms of Roots

Tropical plants of wet forests (none of our weeds are in this category) have aerial roots and absorb water directly from the moist tropical air. Roots of parasitic plants such as the dodder [*Cuscuta gronovii*, p. 138] are, soon after germination of the seed, sent down into the tissues of the host plant and act thereafter as absorbing agents, taking up water and minerals. Some parasitic plants are green and can, therefore, make their own food; they depend on the host plant for water and minerals only. Dodder not only takes water and minerals but also manufactured carbohydrates because it lacks chlorophyll. One green parasitic angiosperm with which you are familiar is mistletoe, formerly a fertility symbol (and still, apparently, partially one at Christmas).

The Flower

The flower is the basic organ upon which classification of the angiosperms is based. Structurally and morphologically, the flower is a *modified branch system*. The portion of the stem on which the flower is borne is called the *receptacle* and is, fundamentally, a compressed region of the stem bearing several crowded nodes and very short internodes.

The lowest organs of the more typical flower are the *sepals*, which collectively are known as the *calyx*. Usually there is one whorl of them. Just above the calyx on the receptacle is the whorl of floral organs known collectively as the *corolla*, each unit of which is a *petal*. (Sepals and petals considered together are known as *perianth parts*.)

Higher up are one or two whorls of *stamens*, each of which is composed of a thinnish *filament* capped by a four-chambered *anther sac*. It is within the latter that pollen is formed. When the sac is mature and the pollen ready to be shed, the sac opens (dehisces) and sheds the pollen grains. (Don't confuse these grains with the fruit, which is called the *grain*. There is really no similarity.) Pollen may be dry or sticky; the dry pollen usually is carried away by the wind, and the sticky awaits an animal pollinator and then sticks to some part of the animal. It is then carried to the next flower and left. Herein lies the reason that goldenrod [*Solidago* species, p. 110, 112] is not the culprit that causes hayfever. This disease obviously can only be caused by a wind-pollinated species of plant. Pollen grains, wind- or animal-carried, contain the fertilization (male) nuclei.

The topmost floral organ is the carpel, which when single is called a *pistil*. The pistil is composed of a basal *ovary* and a thinnish *style*, topped by the *stigma*. Within the ovary chamber are produced one or more ovules. Later the ovary will mature into the *fruit* and the ovules into *seeds*, but only after fertilization has occurred. Pollen grains land (or are deposited on) the stigma.

Pollination of Flowers

The two major pollinating agents of the flowering plants are insects and wind, but there are others. Water effects pollination in species that grow underwater—for example, *Anacharis* (*Elodea*), the waterweed—and many creatures other than insects are involved in pollination—among them slugs, birds, and even monkeys! Of course, man has been consciously pollinating plants since ancient peoples carried the male flowers of the date palm to the female flowers as part of what was very likely a religious ceremony. We still consciously pollinate in our huge genetic programs in order to improve agricultural plants. These programs have been quite successful in creating new varieties and in improving old ones.

Natural animal pollination has resulted in flowers that are larger, more showy, and conspicuous, that are more likely to be fragrant and to have rewards for the visiting animal, such as nectar. Indeed, when we say "flower," we tend to think of big, colorful, attractive corollas. The word "attractive" itself gives the story away; they are attractive to us aesthetically, but to the pollinator in far more practical ways.

Wind pollination, known technically as anemophily (literally, wind-loving), has resulted in quite the opposite kind of blossom. It is not at all necessary to attract the wind, so wind-pollinated flowers do not put on a big show. They are usually small, inconspicuous, nonfragrant, contain no rewards, and have even lost their sepals and petals, which have been selected against during their evolution. Grasses are wind-pollinated.

Pollination and Fertilization

When a pollen grain lands on the stigma of the correct species, it will germinate. A tube grows out of the grain and the fertilization nuclei enter the tube and begin their trip. The tube continues to grow through the stigma, digesting its way through the tissues, down the style, and into the chamber of the ovary. It heads for an ovule, enters it, and deposits the nuclei of fertilization within the egg sac of the ovule where the egg nucleus awaits. After fertilization has been accomplished, a new embryo plant forms by cell division within the tissues of the ovule.

The ovule, once fertilization has occurred, itself undergoes growth and slowly, while the embryo is developing within, becomes a seed. At maturity, the seed contains food tissue that surrounds the embryo plant (or is within its cotyledons).

Not only has the ovule been developing after fertilization, but so has the ovary. The latter structure gradually matures into a *fruit*. The mature fruit may be fleshy or dry. If fleshy, it may contain one or more seeds; a fleshy one-seeded fruit is generally called a *drupe*. The cherry and the peach are examples of drupaceous fruits. If the fleshy fruit contains more than one seed, it is a *berry*. Tomatoes and grapes are berries, and special forms of berries are apples, oranges, and fruits like the cucumber. It is interesting to think that the huge watermelon (fleshy and many-seeded) is a ripened ovary grown to great size.

If a mature fruit is dry, there are two possibilities. It may open or it may not. If it opens, then it may open (dehisce) along one line (suture) and is called a *follicle* (the milkweeds, p. 198, produce follicles), or along two lines and is called a legume (the black locust, *Robinia*, p. 26, is a legume producer), or along many lines, in which case it is called a *capsule* (evening primrose, p. 108, produces capsules).

However, not all dry fruits open and those that do not fall into two major categories: fruits whose single seed is fused to the ovary wall at but one point and fruits whose seed is fused entirely to the ovary wall. If the seed is fused at but one point, the fruit is an *achene*. Composites generally produce achenes. The true fruits of the strawberry are achenes and the strawberry is by no means a berry. The edible portion of the strawberry is the swollen, juicy, and sugary *receptacle*. The tiny,

dark dots on the strawberry are the true fruits, the achenes. To demonstrate to yourself that the seed is attached at but one point in an achene, take a sunflower "seed" (an achene) and longitudinally open it with a razor blade. You will see that the one seed within is attached at only one point.

The second type of dry fruit does not open at maturity. Its contents—one seed—are completely fused to the surrounding ovary wall. An example of this type of fruit, the *caryopsis* or *grain*, is the fruit of the members of the grass family. If you dig a kernel (caryopsis) from an ear of corn (inflorescence—each kernel is the ripened ovary of the pistil of one flower on the ear; the corn silk is a mass of styles and stigmata of these ovaries) and hold it with the point of attachment to the ear down and the white shield-shaped area facing you, you will recognize the major portions of the fruit relatively easily.

The tough outside of this grain is the ripened ovary wall, the white shield-shaped area the single cotyledon that is attached to the embryo corn plant (not visible to your eye at this point). You can make this structure out because it appears as a raised line at the center of the cotyledon. The yellowish material around the white cotyledon and embryo is the food material of the grain.

With a sharp razor blade, try sectioning longitudinally through the grain, passing straight down the embryo. Separate the two equal parts of the grain and look inside. For this you may need a low-power magnifying glass. Within, you can see the future leaves and stem of the embryo maize plant (upper portion of the embryo), the attached single cotyledon pressing against the food material. During germination, this structure releases enzymes that change the stored starch to sugar and then absorb the sugar, passing it to the embryo. The embryo is in a period of crisis during germination; it needs energy for the increased metabolism occurring at this time, and can get this energy only from the sugar stored within the seed since the young plant is unable to make its own food. Below the attachment of the cotyledon is the future root of the new plantlet.

When looking into a grain, you are looking into the source of Western civilization; maize is still, as it was when Columbus arrived in 1492, the basic agricultural grain plant of our hemisphere. Its structure is fairly similar to that of the grains of many noxious grasses of lawn and garden, such as, crabgrass (p. 66).

The Dicotyledonous Seed

While the grain was a fruit we treated as a seed, the dicotyledonous seed of the bean or pea plant is a true seed. Its ovary wall is a pod (legume) that splits open at maturity and permits the ripened ovules (seeds) to fall out.

Soak some lima or kidney beans for about an hour, and when soft, take a bean out of the water and remove the skin on the outside. This skin, the *seed coat*, protects the seed until germination occurs. In some species of plant the seed coat is so tough that it is almost impossible to get water through it. After removing the seed coat, you will see the two *cotyledons* that join each other perfectly, much the way clam shells do..

Insert a fingernail between the cotyledons and open them out. Note their thickness. In the bean the seed food has been absorbed into the cotyledons (it was outside the single cotyledon of the corn grain). When the two cotyledons are opened up, one usually breaks away from its attachment. It was attached to the small embryo plant, which can be seen near the surface of the bean near the top. It is whitish and has two small leaves that make it look like a little feather; hence its name, *plumule*. The cotyledons are attached just below those leaves, and right below their attachment is the *radicle*, or future root.

When germination occurs, the seed leaves (cotyledons), still folded together, will break the soil first, thus protecting the young regular leaves and the shoot apex between them. Once the root is absorbing water and the first regular leaves have turned green and are photosynthesizing, the plant will no longer need to call on the food reserves within the cotyledons. These will then shrivel up and fall away from the plant.

A considerable number of dicot seeds, such as *Ricinus communis* (the castor oil plant) is so structured.. They do not store food within the cotyledons (as we saw in the maize caryopsis) but have a similar structural arrangement. Two cotyledons, their upper surfaces lying closely pressed together and their lower surfaces pressed against the food, act as absorbing agents during germination. The cotyledons slowly slide out of the seed as germination progresses, absorbing until the last moment and then unfolding and turning green in the light. These seeds contain ricin, a very dangerous poison.

THE SPECIES

Prunus

Prunus serotina
▶ Wild black cherry

Wild black cherry is a stately ninety-foot tree with a four foot diameter. Its wood is highly prized in cabinet-making and as a result, few trees attain full maturity.

Why include a tree among the weedy plants? The fruits are beloved of birds, who drop the completely indigestible seeds along fence rows, from telephone wires, etc., where thickets of much-branching young *Prunus serotina* spring up. They are neither attractively shaped nor valuable in this condition.

The bark of the younger branches is reddish-brown, with the horizontal markings common to the genus to which the plum, apricot, cherry, chokecherries, and almond trees belong.

The broadly lance-shaped to oblong leaves are bright green and shiny on their upper surfaces. Their lower sides are lighter green and taper to a point, with small incurving teeth on their margins.

Lovely six-inch racemes of white flowers appear in late May or June. Later green fruits are seen, and still later, the fruits are black and taste a bit bitter.

P. serotina is the most dangerous of the eastern wild cherries. 100 grams of fresh leaves contain ten times the minimum amount of prussic acid (hydrogen cyanide) considered dangerous.

COMMON NAME Wild black cherry

SOME FACTS Native; perennial; propagates by seeds in stones or pits

RANGE Nova Scotia to Florida, west to the Dakotas

HABITAT Woodlands, fence rows, roadsides, waste places

SEASON May–June

Robinia

Robinia pseudoacacia
▶ **Black locust, locust**

Some orphans were playing while fence posts
of black locust logs were being erected. They
began to chew the inner bark of the posts
and became badly poisoned. All fell into a
stupor, and all recovered.

Powdered bark has poisoned a horse
when in an experiment an aqueous extract
representing only 0.1 percent of the horse's
weight was administered. Powdered black
locust bark contains a heat-labile phytotoxin,
and a glycoside has also been found. Fatalities
from bark chewing are rare.

The black locust can grow seventy-five
feet high and has a straight, long, slender, and
coarse-barked trunk. Young branches are
armed with tough spines with stipules, which
may persist for several years.

Alternate, pinnately-compound leaves are
found on the stems and each leaf is composed of seven to nineteen two-inch oval to
elliptical leaflets.

In June the white flowers make their
appearance. The blossoms are very fragrant
and are typical leguminaceous flowers, each
with its standard, wings, and keel.

COMMON NAMES Black locust, locust

SOME FACTS Native; reproduces by seeds,
and roots give rise to new plants

RANGE West, Pennsylvania to Indiana and
Oklahoma; south, to Georgia and Louisiana

HABITAT Roadsides, waste places, open
woods

SEASON June

Ailanthus

Genus

Ailanthus altissima
▶ **Tree-of-heaven**

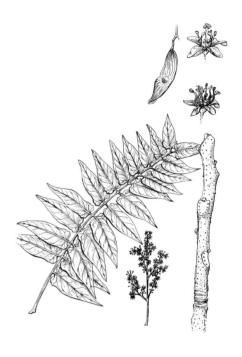

Though *Ailanthus* is in very ill repute, it deserves credit for one thing—it survives handsomely under the most adverse conditions. *Ailanthus* grows from the very cracks of the tenement walls and in garbage- and bottle-filled lots providing a touch of green for people who rarely see any.

The weed tree grows several feet in one year. Large leaves composed of eleven to forty leaflets are alternately arranged on the stout stems. When they fall, they leave large horseshoe-shaped scars on the olive-tan stems. Each leaflet of a leaf of *Ailanthus* is narrow-oblong and near the base.

Ailanthus are either male or female. The male's flowers give off a vile odor, leading to the name stinkweed. The greenish-white female flowers produce fruits called samaras. Each samara contains a single seed with wings that are borne in large clumps, pale gray-yellow in color, not at all unattractive. Each samara is twisted, much like a propeller, and when it falls from the tree, the wind spins and carries it some distance from the tree that bore it.

COMMON NAMES Tree-of-heaven, stinkweed, tree-that-grows-in-Brooklyn

SOME FACTS From North China; perennial; reproduces by seeds

RANGE North temperate, oriental

HABITAT Vacant lots, roadsides, open woodlands

SEASON June–July

Family

Berberis

Berberis thunberghii
▸ **Japanese barberry**

Japanese barberry was introduced later than
European barberry and is much smaller, growing to six feet at the tallest, but is usually seen
two or three feet high. It is a useful hedge
plant because of its rapid growth and its
unbranched spines, about a third as long as
those of its cousin, but still sharp.

Its leaves are short and delicate in appearance, narrowing to a short petiole. Usually
there is a purple cast to the leaves because of
the presence of anthocyanin pigments in the
sap. They may turn bright red in the fall.

Yellowish sweet-scented flowers appear in
May or early June. Eventually berries that will
turn red will take their place. Because they are
relatively dry they are not used in jams and
jellies, but they are eaten by birds who then
distribute them across the countryside.

Berberis vulgaris

On *B . thunberghii*, the Japanese barberry, the
spines are not three-pronged but much
smaller. Common barberry also produces
long clusters of blossoms, while its more
recently introduced Japanese cousin usually
produces two short clusters in the axils of
leaves. The common barberry has heavily
perfumed, small yellow six-parted flowers,
appearing in drooping racemes composed of
ten to twenty flowers each. Red berries, popular for making preserves, appear later in the
season and contain from one to a few seeds.

COMMON NAME Japanese barberry

SOME FACTS Introduced from Japan;
perennial; reproduces by seeds and
creeping roots

RANGE Northeastern states of the
United States

HABITAT Gardens, waste places, borders
of woods

SEASON May–June

Rubus allegheniensis
▶ **Common blackberry, bramble**

Many know the wild blackberry for two reasons: its fruits are delightful and its thorns are dreadful. This very unpopular weed has long, erect, stout canes covered with tough prickles. Weight causes longer stems to arch over. Where their tips touch the ground, new plants may appear.

Bramble is a relative of plants that produce edible fruits, such as the raspberry and wineberry. The blackberry and its cousins are members of the great Rose Order.

The blackberry's compound leaves are eight inches long at full growth and are divided into five leaflets, all arising from one point at the far end of the petiole. Such a leaf is said to be palmately compound. Smaller leaves on flowering branches are usually composed of but three leaflets. All leaflets are finely toothed.

The delicious blackberries, which American Indians used as food, are preceded by white flowers produced in racemes. Each blossom is about three-quarters of an inch in diameter.

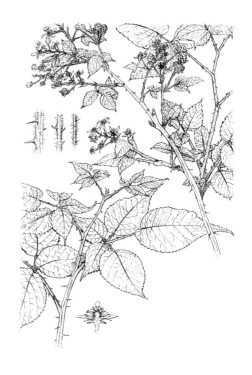

COMMON NAMES Common blackberry, bramble

SOME FACTS Native; perennial; reproduces by seeds and rootstocks (runners)

RANGE Nova Scotia to Quebec and south to edge of mountains in North Carolina and Tennessee

HABITAT Open places, edges of woods

SEASON May–June

Family

Lonicera
Genus

Lonicera japonica
▸ **Japanese honeysuckle**

Under control, the climbing vine called Japanese honeysuckle is lovely to look at. Once escaped, it will cover large areas of low-growing plants with its spreading stems, obliterating everything by climbing and shading saplings.

It travels through forests by runners and by seeds. The latter is successful because birds scatter the seeds far from the mother plant. The runner makes this plant diabolical in its speed of travel. Runners average twenty-five feet in one year, and one sprout has covered more than forty-five feet in one season!

The hairy leaves are ovate-oblong and simple, with even margins, remain green throughout the year and are ready for an early start in the spring. The leaves lower down the stems have short petioles. Those near the top are sessile and sometimes each pair of leaves is fused together.

The plant's yellow or white blooms occur in pairs, each having five petals fused to form a tube and a two-lipped corolla. Purplish berries result from fertilization.

COMMON NAME Japanese honeysuckle

SOME FACTS Introduced from Asia; perennial; reproduces by seeds and by creeping stems

RANGE Massachusetts to Indiana, south to Florida and Texas

HABITAT Fields, thickets, walls, gardens, waste places

SEASON June–July

Family

Convolvulus
Genus

Convolvulus arvense
▸ **Wild morning glory, hedge bindweed**

Hedge bindweed or wild morning glory is a rather unwanted plant because of its rapid, widespread growth covering over more valuable species. Still, it does produce large bell-shaped flowers and long trailing or twining stems from three to ten feet in length; these are smooth or hairy. *C. arvense's* brittle, cord-like roots rapidly spreads the plant.

Smooth leaves, triangular-ovate to halberd-shaped, with divergent basal lobes, are alternately distributed along the twining stems. Their petioles are usually shorter than the leaf.

Individual flowers one to three inches across, pink or white and bell- or trumpet-shaped, arise in the axils of the leaves on peduncles (bases) four to six inches in length. Two heart-shaped bracts are found at the base of the flower hiding the five petals, and later they enfold the four-seeded capsule. Bracts are not seen on *C. arvense*, the field bindweed. Also field bindweed's flowers are smaller, one inch across at best, and several appear in each axil.

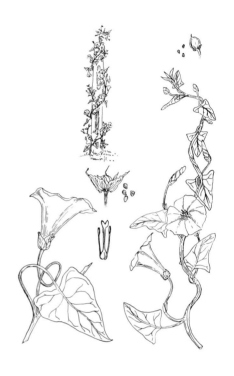

COMMON NAMES Wild morning glory, hedge bindweed, devil's vine, great bindweed, bracted bindweed, hedge lily

SOME FACTS Native to America; perennial; propagates by seeds and vigorous rootstocks

RANGE Found throughout northeastern and north central United States

HABITAT Fence rows, meadows, cultivated fields, waste places

SEASON June–August

Family

Smilax glauca
▶ Greenbrier, catbrier

Catbrier is very unpopular with nature lovers. Getting into a patch of catbrier can be very disconcerting because the stems are armed and can inflict severe damage on the skin and clothing.

With its round stems and stout curved prickles it does not seem natural to place greenbrier among the monocots . Leaves that are broadly ovate sit on petioles, at the base of which is found a persistent tendril that remains on the vine after the leaves have fallen. Small umbels of six-parted yellow-white flowers emerge from the axis of some of the leaves.

An edible jelly can be made from the roots. Dissolved in water, it makes a relatively palatable drink. The American Indians used these roots after this fashion. True sarsparilla is made from the roots of a tropical *Smilax*. The young shoots of *Smilax* are tasty in a salad or cooked.

A close relative is S. *herbacea*, which has unarmed stems but armed flowers, which produce a disgusting odor, giving it the common name carrion flower.

COMMON NAMES Greenbrier, catbrier, false sarsparilla

SOME FACTS Native; perennial; reproduces by seeds and tubers

RANGE Common along Atlantic and Gulf coasts

HABITAT Open woods, thickets, roadsides

SEASON May–June

Vitis
Genus

Vitis labrusca
▶ **Fox grape, wild grape**

V. labrusca, wild or fox grape, has a stem, characterized by its shreddy outer tissues and a diaphragmed pith, lacks much of the supportive, mechanical tissues possessed by the plants on which it so often is found growing. It clutches its support by means of tendrils. Despite the lack of supportive tissue, one shoot may reach one hundred feet in length.

Fox grape can be parasitic because it covers the foliage of its living support, thus preventing sunlight from reaching the host's leaves. It prohibits its living support from making its own food and the support will die. Dense tangles of grape may grow up and over the leaf cover of shrubs and lower trees, covering acres of herbs, shrubs, and trees. In the wild, whole areas may be threatened.

The large, eight-inch leaves are as wide as they are long and are shallowly three-lobed. On their lower surfaces the hairs are so dense that the surface itself is invisible. The serrations along the edges are various sized. These large leaves are opposite and simple.

The flowers, borne in racemelike panicles, are small and relatively fragrant. Each contains stamens which are found exactly opposite their petals. The fruit which follows pollination is popularly called a grape (from which wine is made), but is botanically a berry; that is, it is a fleshy fruit which contains more than one seed.

COMMON NAMES Fox grape, wild grape

SOME FACTS Native; perennial; reproduces by seeds and rhizome

RANGE Maine to southern Michigan, south to South Carolina and Tennessee

HABITAT Roadsides, thickets

SEASON May–June

Family

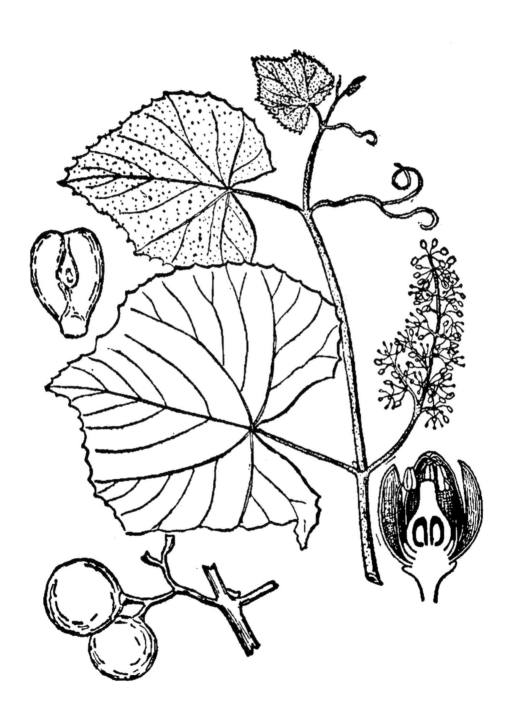

Echinocystis

Echinocystis lobata
▶ Wild cucumber; balsam apple

Balsam apple climbing unsupervised can act as a parasite covering the surfaces of the supporting plant, denying it sunlight, and thus stunt or kill. it.

Under control it is an attractive plant grown for shade or to cover a fence or tree stump. Its fifteen- to thirty-foot stems are smooth except at the nodes, where a few hairs may be seen. Its alternate leaves, generally orbicular in outline, are palmately lobed and veined. Five is the most frequent number of leaf lobes. A sinus is seen where the relatively long petiole meets the leaf. Opposite the leaf is found a three-forked tendril.

In the axil of the tendril appear the white, fragrant, five- or six-parted staminate and pistillate flowers. The staminate flowers attract attention in prominent racemes, while the pistillate flowers, inconspicuously emerge from the same axis. The two-inch ovoid fruit open and the seeds are ejected. The old, inner, fibrous-netted dried fruit's spines can jab painfully.

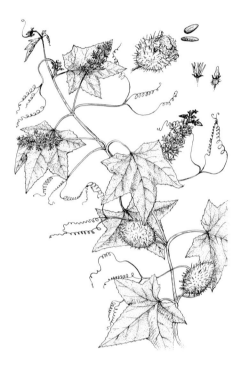

COMMON NAMES Wild cucumber, balsam apple, mock apple, four-seeded bur cucumber

SOME FACTS Native to the United States; annual; reproduces by seeds

RANGE Widespread in northeastern U.S., and westward to Saskatchewan and Texas

HABITAT Fence rows, damp rich soil, thickets

SEASON July–September

Rhus
Genus

Rhus radicans
▶ Poison ivy, markweed

Few plants are more disliked or feared than *Rhus radicans*, also known as poison ivy and markweed.

The leaf is composed of three shiny leaflets (note leaflets, not leaves), and the old adage, "Leaflets three, let it be," is one worth heeding. The vine is often confused with the perfectly charming and innocuous Virginia creeper (*Parthenocissus quinquefolia*), which has five leaflets per leaf and should not be destroyed; therefore, it may be wise to remember a new adage: "Leaflets five, let it survive" (p. 48).

All parts of the three-leaved ivy (observe that this common name contains a botanical error) are poisonous except for the pollen grains. Tearing and breaking the stems even in midwinter is dangerous to man and other closely related primates, though animals situated lower on the evolutionary ladder seem to be relatively safe. Birds eat the small white berries without ill effects—to themselves (later they "plant" the seeds far from the original vine and thus spread the plant).

It is said that the juices from jewelweed (*Impatiens biflora*, p. 136) will prevent poisoning by this plant or halt its progress in one who has become contaminated. Indeed, some drugstore preparations contain substances from jewelweed. Taking a bath in water to which crushed, boiled jewelweed leaves have been added is said to give excellent protection.

COMMON NAMES Poison ivy, markweed

SOME FACTS Native; perennial; reproduces by seeds

RANGE Nova Scotia to British Columbia, south to Florida, Arkansas, and Utah

HABITAT Roadsides with stone walls, banks and waste places, climbing trees

SEASON Late May–July

Family

Vicia

Vicia cracca
▶ Wild vetch, blue vetch

The lovely one-sided axillary racemes of the flowers of the blue vetch enliven the banks along the sides of highways. Rain leaches nitrates from roadside slopes and the root tubercles on wild vetch make nitrates, permitting this weed to grow. Because of its tough creeping roots, it is difficult to remove from areas where it has become undesirable.

A slender climbing or spreading stem may sport hairs. The tendrils at the tips of its pinnately-compound leaves help tufted vetch to climb—on other plants and smother them—or spread on the ground and shade out grass or other desirable plants under it. Each leaf is composed of eight to twelve pairs of thin, oblong-lanceolate leaflets, each of which ends in an abrupt tiny point, the leaf itself ending in a tendril. The whole plant is a soft olive green in color.

The numerous flowers are approximately as long as a leaflet, about one-half inch, and hang bent outward. The pods (legumes) that appear after fertilization are approximately an inch in length.

COMMON NAMES Wild vetch, tufted vetch, blue vetch, cow vetch, bird vetch, cat peas, titters, tine grass

SOME FACTS Introduced from Eurasia; perennial; reproduces by seeds and rootstocks

RANGE Northeastern United States and eastern Canada, also found on Pacific Coast

HABITAT Fields, meadows, waste places, roadsides, especially along banks

SEASON June–July

Parthenocissus

Genus

Parthenocissus quinquefolia
▶ Virginia creeper, American ivy

Virginia creeper is not a weed but has been included to protect it. Virginia creeper has been burned, hacked, and uprooted because most people are unable to tell the difference between it and poison ivy.

Both poison and American ivy have compound leaves, but the true similarity stops there. Virginia creeper has long-petioled, palmately compound leaves composed of five leaflets each a dull, darkish green, elliptical to obovate and serrated beyond its middle. Poison ivy's lighter green (almost yellow) and shiny-compound leaf is composed of three leaflets.

Later in the season, white berries are found in clusters on poison ivy while Virginia creeper is sporting berries that are nearly black. Each one of these berries contains one to four seeds and was produced by a flower whose parts were in multiples of five. The flowers are produced in panicles that are usually longer than they are wide; the panicles are produced opposite the leaves.

It is not selective of its companions and may be found climbing in the same place or on the same wall or tree as *Rhus radicans*. Both climb with the support of special organs; poison ivy using modified roots and Virginia creeper by means of adhesive discs at the end of much-branched tendrils.

COMMON NAMES Virginia creeper, woodbine, American ivy, five-leaved ivy

SOME FACTS Native; perennial; propagates by seeds

RANGE Maine to Ontario and south to Florida and Texas

HABITAT Moist soil, rich woods in disturbed areas, along fencerows and walls

SEASON June

Family

Typha
Genus

Typha angustifolia
▶ **Cattail**

Cattail may be wide or narrow-leaved, the narrow-leaved form being considered a separate species, *Typha angustifolia*. The two species look alike and grow under similar conditions.

Cattail grows in marshes with its roots under water. Stands of cattail composed of many plants are usually seen because of the rapid growth of its rhizome. From these arise one to nine-foot-long culms with long, half-inch-wide, clasping, linearly sword-shaped leaves.

The "punk" is a spike of mature fruits and their attendant hairs and was, before fertilization, the spike of female flowers. The male spike is borne immediately above the female and these flowers drop from the culm after their pollen has been shed.

The green spikes, when cooked with salty water, make a tasty vegetable, and the pollen, mixed with wheat flour, can be used in making pancakes. Russians living along the Don love the young shoots, cooking them like asparagus or eating them raw.

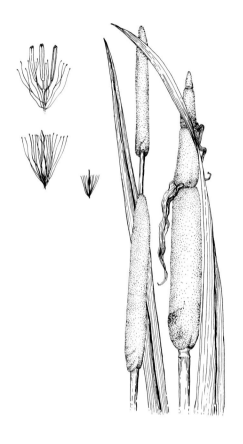

COMMON NAME Cattail (bulrush or reed-mace in England)

SOME FACTS Native; perennial; reproduces by seed and rhizomes

RANGE Throughout North America (except in extreme north)

HABITAT Marshes, very wet soil, drainage ditches along roads

SEASON July–August

Family

Cyperus
Genus

Cyperus strigosus
▸ Straw-colored cyperus, false nutsedge

Sedges are first cousins to the grasses but may be separated from them in a number of ways. The stem of a sedge is triangular and solid throughout its length. The stem of a grass is round, solid at the nodes and hollow in between.

The sedge's distribution of leaves is three-ranked (each leaf is one-third the circumference of the stem away from its higher neighbor or its lower neighbor) and that of the grasses two-ranked (each leaf is separated from the others by one-half the circumference of the stem).

Grass leaves are flat, while sedge leaves are grooved or V-shaped. Few sedges become weedy, whereas many grasses are included in this designation.

Arising from a hard cormlike base that is slightly swollen, *C. strigosus* stands two to three feet high and is found on poorly drained land. It is related to the hummock sedge and the bulrushes.

The flat spikes of wind-pollinated flowers are borne in either simple or compound umbels surrounded by an involucre of three flat, smooth leaves.

COMMON NAMES Straw-colored cyperus, false nutsedge, lank galingale

SOME FACTS Native; perennial; propagates by seeds and by tubers; cormlike

RANGE Widespread in eastern U.S., also found on Pacific Coast

HABITAT Along streams, in damp meadows, poorly drained fields, ditches

SEASON July–September

Family

Allium

Allium vineale
▶ Wild garlic, wild onion

Wild garlic is a close relative of the table onion and has an odor that brings tears to the eyes.

The slender, pointed, hollow, dark green leaves, round in cross section, emerge from an underground bulb. Wild onion produces two kinds of bulbs: soft bulbs, which start growth during their first fall; and hard bulbs, which are dormant in winter and then germinate during their first spring. Hard bulbs may remain dormant for longer than one year.

From late May to late June the flowering stalk of this member of the Liliales may be seen to arise from the center of the leaf cluster. Small pink-purplish six-parted flowers are produced in umbels. These flowers are replaced by small bulblets, each of which is tipped by a slender filament or tail. One flower seed head may contain thirty to one hundred bulblets, each approximately the size of a wheat grain. Indeed, wheat grain is occasionally adulterated by these bulblets (called cloves by some), and bread made from wheat ground with these bulblets will be unmarketable.

COMMON NAMES Wild garlic, wild onion, field garlic, crow garlic

SOME FACTS Introduced from Europe; perennial; reproduces by bulbs and bulblets formed along flowers, rarely by seeds

RANGE Massachusetts to South Carolina to Mississippi

HABITAT Fields, meadows, pastures (prefers sandy loam)

SEASON May–June

Juncus
Genus

Juncus tenuis
▶ Path rush, wire grass, yard rush, field rush

Walk down a dew-laden path early some morning and then check your shoes. Chances are that a sizable number of small orange-brown gelatinous seeds will have stuck to their sides. These are the seeds of the path rush, a plant found abundantly along the paths man makes through fields.

Someone not trained in botany might easily think that *Juncus* is a member of the grass family; however, it is not a grass. Its flowers have sepals and petals, and though both are harsh and strawlike, they are real sepals and petals. Furthermore, the fruit is a capsule filled with small seeds. A proper grass produces a grain (caryopsis) that contains but one seed.

This slender rush seems happy on wet or dry soil. Its thickly tufted stems may be six inches to two feet tall and are very thin, round, and wiry enough to spring back quickly after being trod upon. Its complete flowers appear in irregular clusters between two long, flattened leaves at the top of the stem. Each is composed of three green sepals and three green petals, lanceolate and sharp-pointed.

COMMON NAMES Path rush, wire grass, yard rush, field rush, poverty rush, North American rush (by the British)

SOME FACTS Native to United States; perennial; reproduces by seeds

RANGE Widespread in United States and Canada

HABITAT Waste places, along paths in fields, roadsides

SEASON June–August

Family

Phleum

Genus

Phleum pratense
▶ **Timothy, Herd's grass**

Timothy may be the most important pasture grass cultivated in the United States. It has escaped from cultivation and become a weed. This tall grass grows well in the cool, humid climes of the Northeast and south into the cotton belt. It is also found as far west as Puget Sound, along the coastal area of the Pacific Northwest, and in the valleys of the Rocky Mountains.

Once timothy has put out its crowded terminal green spike (a tightly contracted panicle), it is immediately recognizable. Its twenty- to forty-inch-long stems emerge from a swollen or bulblike base that forms large clumps. The "bulb" or "corm" at the base (and there may be two) is actually a swollen or thickened internode.

Each "bulb" is annual in duration; it forms during the summer and dies the next year when the seed matures. Timothy, used as pasture grass, should be cut in early bloom because the food value of the grass decreases and its fiber content increases as the season advances.

COMMON NAMES Timothy, Herd's grass

SOME FACTS Introduced from Europe; perennial; propagates by seeds; most important hay grass in the United States

RANGE Temperate regions of both hemispheres

HABITAT Humid regions of northeastern United States, open lots; likes clay loams and lightly textured sandy soils

SEASON June–July

Family

Setaria
Genus

Setaria viridis
▶ Yellow foxtail grass, wild millet

Both this plant and species of Panicum (see pp. 70–73) are closely related to foxtail millet (*Setaria italica*), also known as summer grass, golden foxtail, and pussy grass. Millets are a poor man's cereal, as well as food for birds and turkeys.

Wild millet may attain a height of four feet, but plants one to three feet high are common. Yellow foxtail grass will flower only a few inches if it has been repeatedly cut before flowering time. Its stems branch at their bases and bear three- to six-inch-long flat, smooth, linear-lanceolate leaves that hang with a twist. They are one-half inch wide.

The spikelets of the one- to four-inch-long spike contain one flower each and are closely packed together. Each seed is subtended by a cluster of yellowish-brown barbed bristles that are much longer than the seed. As a result, the whole inflorescence looks like a bottle brush or "foxtail." When the inflorescence has aged and dried, it may be dangerous to cattle, producing oral disturbances and infections.

COMMON NAMES Yellow foxtail grass, summer grass, golden foxtail, wild millet, pussy grass

SOME FACTS Introduced from Europe; annual; reproduces by seeds

RANGE Common throughout North America

HABITAT Waste places, rich soils, pastures, cultivated ground; likes just about any soil

SEASON July–September

Family

Lolium

Genus

Lolium perenne
▶ Perennial ryegrass, common darnel

Perennial ryegrass (or raygrass) was originally grown in pure stands as a forage grass and is still a popular forage grass in Europe, where the cool moist summers are ideal for it.

The erect culms of common darnel stand ten to thirty inches high and are smooth. Leaves are flat and smooth. The mature plant is reddish near its base.

A three- to eight-inch terminal spike of flowers is produced, on which the flattened spikelets are pressed closely against a concave space in the rachis. Each spikelet is enclosed in but a single glume and each is composed of five to ten flowers. No awns are seen on the lemmas or on the seeds.

Common darnel's close relative, poison darnel (*Lolium temulentum*), was known to one of the authors of the Bible, who was probably referring to this plant when he wrote of "tares." Theophrastus, colleague of Aristotle, knew this plant. The poison, lacking in perennial ryegrass, is produced by a fungus that attacks the poison darnel.

COMMON NAMES Perennial ryegrass, raygrass, common darnel, English ryegrass

SOME FACTS Introduced from Europe; perennial; reproduces by seeds

RANGE Throughout northern United States

HABITAT Fields, meadows, roadsides, pastures, lawns

SEASON June–July

Family

Andropogon
Genus

Andropogon virginicus
▶ Little blue stem, broombeard grass

The translated scientific name of this grass gives a good description of some of its features. In Greek andros means "man" and pogon "beard." The inflorescences of little blue stem contain filmy hairs, which give them a bearded appearance, while the overall aspect of the plant is broomlike.

Standing one to three feet tall, the flat, erect stems occur in tufts. The leaves of this native bunch grass are, at first, light green, but later became reddish. Two types of spikelets are found on little blue stem: a sessile fertile one and a stalked sterile one. The latter bears the numerous filmy hairs that give the inflorescence its bearded appearance.

Little blue stem is highly prized as a forage grass, along with its cousin *Andropogon gerardi*, which is known as big blue stem. Both form the most prevalent constituents of wild hay in the prairie states. When mature, both grasses are less palatable. Broombeard grass thrives on a wide range of soils, and since it is drought-resistant, it is useful in erosion control.

COMMON NAMES Little blue stem, broombeard grass, wolf grass, poverty grass

SOME FACTS Native to the United States; perennial; reproduces by seeds

RANGE Widespread in the U.S., especially in prairie states

HABITAT Waste ground, dry fields, pastures, especially in sandy soil

SEASON July–September

Family

Digitaria

Genus

Digitaria sanguinalis
▸ Crabgrass, fingergrass

Fortunately, crabgrass, also called crowfoot grass and pigeon grass, is annual, but unfortunately, the seeds it forms in only one growing season are long-lived, and wherever the stems touch the ground, the nodes root.

In the South it is a blessing to the farmer who wants a rapidly-growing hay plant with pasturage possibilities after the hay has been collected. Strawberry growers let it fill in the spaces between the rows of berry plants where otherwise they would have to scatter straw.

Crabgrass can get as long as four feet, though is usually shorter. Each leaf is three to six inches long and one-fourth inch wide.

Three to ten spikes of tiny flowers top the culmand, each two to five inches. The scientific name of fingergrass (*Digitari*) comes from the handlike appearance of the stems. Since they turn bright red, blood (*sanguinalis*) is suggested. It germinates from seed on the lawn between midspring and late summer and is killed by the first frost.

Crabgrass can take neither the heat and sunlight nor shading and is caused by improper mowing.

COMMON NAMES Crabgrass, fingergrass, Polish millet, crowfoot grass, pigeon grass

SOME FACTS Introduced from Europe; annual; propagates by seeds and by rooting nodes

RANGE Worldwide, widespread throughout America, very noxious in Northeast

HABITAT Waste places, bare patches in lawns, along ditches, in cultivated fields

SEASON July–September

Family

Eleusine

Genus

Eleusine indica
▶ Goose grass, wire grass

The lawn pest goose grass is often confused with crabgrass. The sessile spikelets, in two rows on two, are on three to eight branches that are digitately arranged at the apex of the culm. This corresponds, though only superficially, with the fruiting structure of crabgrass (see p. 66).

Below the digitate rows of spikelets there may be one or two branches with spikelets. The stem or culm of this low-spreading annual grass is thick, flat, and weak, and often prostrate. In the northern sections of the United States yard grass is usually relatively short, but in the South it may attain a height of two feet. Its leaves have loose, overlapping, flattened sheaths and flat, pale green blades.

It is cultivated in India (hence *indica*) for grain and its seeds were used in poor sections of Europe as a flour source. Its use as a grain source in early times is suggested by its generic name Eleusine, taken from the Greek city *Eleusis* in which a temple of the protectoress of grain was found.

COMMON NAMES Goose grass, yard grass, wire grass, crowfoot grass

SOME FACTS Introduced to United States from warmer parts of Asia; annual; propagates by seeds

RANGE Widespread in the United States, especially in the South

HABITAT Lawns, roadsides, waste places, yards

SEASON June–September

Family

Panicum
Genus

Panicum clandestinum
▸ Panic grass

With its wide, diverging leaves and sturdy culms, its obvious parallel venation of leaves that clearly clasp the stems, panic grass, also known as deer tongue grass, is monocot. Its grass nature is not announced as loudly.

Adding to the difficulty is the fact that its inflorescences are clandestinely produced, that is, hidden from the eye until late in the season when the mature seeds are exposed in panicles. No one could confuse the involucre, if visible, with that of a lily-type flower.

Deer-tongue grass' flowers remain hidden within the clasping leaf bases of the upper leaves, emerging after seed set has occurred. These leaves have overlapping leaf sheaths, while the culm has very much shortened internodes. The much-reduced flowers are clandestinely pollinated, that is, self-pollinated while hidden within the leaf sheaths.

This species of grass is related to *P. milia-ceous*, known to us as millet, thought to be the very first of the cultivated grains.

COMMON NAMES Panic grass, deer-tongue grass

SOME FACTS Native of the United States; perennial; reproduces by seeds and by rhizomes

RANGE Quebec and northern United States to Michigan, Missouri, and Oklahoma, and south to Florida and Texas

HABITAT Moist woods, thickets, along edges of open areas

SEASON June–July

Family

Panicum

Genus

Panicum virgatum
▶ Switch grass

Switch grass is a vigorous, native, perennial, sod-forming grass that occurs throughout most of the United States. It is most abundant and important as a forage and pasture grass in the central and southern parts of the Great Plains.

Its smooth stems grow to five feet under good conditions and are often glaucous. The leaf blades may be a foot long and one-half inch wide. The leaves, too, are smooth and often bluish-green in color. While their surfaces are smooth and flat, their margins are slightly rough to the touch.

Erect six- to twenty-inch-long panicles that are spreading and pyramidal are found on this grass. The spikelets are one-seeded and ovate.

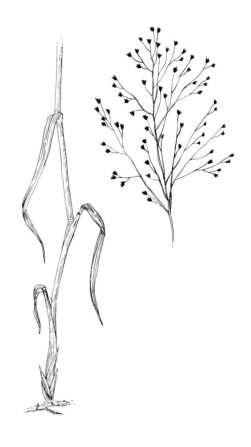

COMMON NAME Switch grass

SOME FACTS Native to the United States; perennial; reproduces by seeds and rootstocks

RANGE Widespread from Maine to Manitoba and south to Florida and Mexico

HABITAT Sandy soil, salt marshes along the coast, along stream banks, in low meadows

SEASON August–September

Family

Bromus

Bromus tectorum
▶ **Early chess, downy bromegrass, slender chess, downy chess, cheatgrass**

Downy bromegrass is a beautiful sight at maturity. The drooping, dense panicles of long slender spikelets look like purple plumes as they nod in the wind at the tops of the one- to two-foot-long stems. Its leaf blades and sheaths are very pubescent.

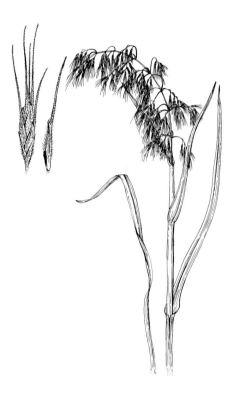

COMMON NAMES Early chess, downy bromegrass, slender chess, downy chess, cheatgrass

SOME FACTS Annual or winter annual; native of Europe; reproduces by seeds

RANGE Widespread in United States except in the Southeast

HABITAT Roadsides, waste places, fields, especially on dry, sandy, or gravelly soils

SEASON May–July

Echinochloa
Genus

Echinochloa crusgalli
▶ Barnyard grass

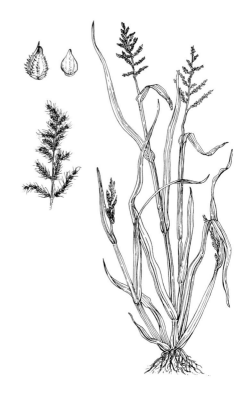

Few people see this weed where it used to be, the barnyard. It invades the garden and its seeds survive for nine years in the soil.

Barnyard grass, also known as cockspur grass and cocksfoot panicum, can be fed to cattle. Arizona and southern California Indians have used the seed for cattle feed and it is eaten as a famine food by the Chinese.

Echinochloa is a tall plant with stout three-foot-tall culms that may be somewhat decumbent. It has long leaves with pale midribs. Its inflorescences makes it easy to identify. It has large four-inch-long panicles that are easily spotted because the branches of this spreading panicle are fringelike and bear long bristles. Each bristle is a barbed awn associated with a flower. Its generic name comes from this bristly spreading panicle at the top: *Echino-chloa* means "green hedgehog" and *crusgalli* means "cock's foot." Sometimes the inflorescence is a deep purple rather than a pale yellow.

Barnyard grass is bothersome in rice and carrot fields in and fields with other root crops.

COMMON NAMES Barnyard grass, cockspur grass, water grass, cocksfoot panicum

SOME FACTS Coarse annual; from Eurasia

RANGE Cosmopolitan

HABITAT Moist ditches, manured soils, cultivated fields, gardens, rice fields

SEASON July–September

Family

Dactylis
Genus

Dactylis glomerata
▶ Orchard grass, cocksfoot

Orchard grass is one of the first grasses to grow in the spring, and among the last to give in to the frosts of fall. This larger member of our grass family was introduced to America in 1760 as a desirable pasture grass. It had been cultivated for centuries in Europe as a meadow and pasture grass. It is not seriously bothered by diseases.

Cocksfoot is a tall, erect grass that grows as high as three or four feet. It grows in great clumps rather than in dense sods. This is because of its failure—Zeus be thanked!—to produce rhizomes. The large tussocks formed from the densely clustered culms are topped in height only by the large dactyloid inflorescence. The densely crowded spikelets are in irregular clusters at the end of the branches of the large, tall panicle.

COMMON NAMES Orchard grass, cocksfoot

SOME FACTS Introduced from Europe; perennial; reproduces by seeds

RANGE Common throughout North America except in desert or arctic regions

HABITAT Open fields, pastures, lawns, gardens, roadsides

SEASON July–August

Family

Phragmites

Phragmites communis
▸ Reed grass, cane grass

It is hard to imagine that anyone with an interest in plant life has not seen *Phragmites*, one of the tallest grasses in our country. Its culms attain a height of six to nine feet under marshy conditions. Topped by a puff of delicate flowers, these culms are often gathered and placed in vases for their aesthetic qualities. The nodding heads of flowers are a conspicuous dull purple.

The generic name of this plant comes from the Greek *phragma*, which means a "fence" or "screen," and the impression of a reed screen is definitely produced by the tall, thin stems, which resemble slim bamboo stalks. The erect stems arise from underground creeping rhizomes.

Reed grass' older stems have been used to thatch roofs, and the young shoots may be either boiled and cooked as asparagus is, or pickled. The rhizome from which the aerial shoots arise can be boiled and eaten just like a potato.

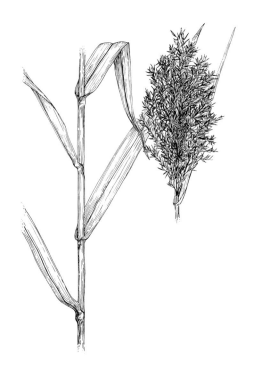

COMMON NAMES Reed grass, cane grass

SOME FACTS European; perennial; rarely propagates by seeds

RANGE Nova Scotia to British Columbia, then south through the U.S.; also in South America and Australia

HABITAT Drainage channels, shallow water, marshy wet shores, along ditches (next to roadways)

SEASON June–July

Commelina
Genus

Commelina communis
▶ Dayflower

The dayflower is quite attractive, but can take over a shady section of the garden so rapidly that other plants may not have a chance to survive. The speed of this weed comes from its creeping stem, which roots rapidly at its swollen nodes.

One of our very few weeds from Asia, its two-to-four inch leaves are alternate, simple, and entire, with the parallel veins expected in a monocot. Commelina is a member of the spiderwort family.

The dayflower has two lateral bright blue petals and a very much smaller and paler (or white) third petal beneath the two lateral ones.

There are also six stamens, only three of which are fertile (three are fused pistils).

COMMON NAME Dayflower

SOME FACTS Native of Asia; annual; reproduces by seeds and by creeping stems

RANGE Massachusetts to Florida, west to Kansas and Texas

HABITAT Gardens, neglected fields, waste-rich soil

SEASON July–September

Family

Plantago lanceolata
▶ Ribgrass, English plantain

Ribgrass plant, also known as buckhorn, ribwort, and ripple grass, grows in open places. The dark green, thin, lanceolate leaves (hence its specific epithet) are deeply ribbed by what seem to be parallel veins, but do not make the erroneous assumption that ribgrass is a monocot. It is a member of the dicots. This crown of lanceolate, ribbed leaves sits on a very short stem (not really observable from above) from which emerge many fibrous roots.

The flowering scapes of English plantain are different from those of the broad-leaved plantain (*P. major*, see p.85) in that the flowering portion of the scape is contracted into a short, dense spike which sits on a long thin, angled stalk. The spikes flower from the bottom, as do those of the broad-leaved relative. The stamens, sticking out from the dense greenish-brown spikes, look like pins in a pincushion.

COMMON NAMES English plantain, ribgrass, buckhorn, ribwort, narrow-leaved plantain, ripple grass, blackjacks

SOME FACTS Introduced from Europe; perennial; propagates by seeds

RANGE Throughout the United States and Canada

HABITAT Lawns, roadsides, waste places

SEASON May–October

Plantago
Genus

Plantago major
▸ **Common plantain, greater plantain, dooryard plantain**

No unattended lawn is safe from the depredations of common plantain, whose rosettes are ellipsoidal, often spoon-shaped, deeply veined leaves that quickly shade out more desirable lawn grasses. Numerous seeds are produced within each tiny capsule and there are many capsules formed along the three- to twelve-inch-long, rattail-shaped spikes of flowers that arise from the center of each rosette. A single, tiny flower is composed of four sepals, four petals, and four stamens with long filaments that stick out of the flower. The capsules, formed after fertilization of the flower, open transversely by a lid.

No weedy species of plantain are put to use by man, but the seeds of the very closely related *P. psyillum*, *P. ovata*, and *P. indica* are used to overcome the sluggishness of man's large colon. John Gerard, the ancient herbalist, has something to say about the usefulness of *Plantago*: "The juice dropped in the eies colles the heate and inflammation thereof.

I find in antient writers many good morrowes, which I thinke not meet to bring into your memorie againe; as, That three roots will cure one griefe, foure another disease, six hanged about the necke are good for another malady, etc., all of which are but ridiculous toyes." I think he has something there.

COMMON NAMES Common plantain, greater plantain, dooryard plantain, white man's foot, broad-leaved plantain

SOME FACTS Introduced from Europe; perennial; reproduces by seeds

RANGE Throughout United States and Canada

SEASON May–September

Chenopodium

Chenopodium album
▶ Mealweed, white goosefoot

Mealweed, also known as lamb's-quarters and white goosefoot, grows very rapidly under good conditions and may attain a height of six feet. Its sturdy stem is ridged and grooved and usually much branched. Stems of older specimens may be striped with reddish purple.

Near the base of the stem are seen alternate petiolate leaves that are rhombic-ovate or goosefoot-shaped. Higher up on the stem the leaves are narrower and lanceolate, and at the top of the plant they become linear and sessile. The leaves are dark green when viewed from above and whitish or light gray-green and mealy when viewed from below.

Small green flowers are crowded on irregularly spiked clusters in panicles found within the axils of leaves and at the tip of the plant. The five lobes of the mature calyx enfold the lens-shaped dark-colored seeds, each of which has a marginal notch. These seeds have been shown to be very long-lived.

COMMON NAMES White goosefoot, lamb's-quarters, pigweed, fat hen, meal-weed, frost-blite, bacon weed

SOME FACTS Introduced from Eurasia; annual; reproduces by seeds

RANGE Common throughout the United States

HABITAT Waste ground, cultivated fields, gardens

SEASON June–September

Ambrosia

Ambrosia trifida
▸ **Wild hemp, giant ragweed, bitterweed**

Giant ragweed, also known as wild hemp and bitterweed, is a member of the tribe of composites and grows to a great height. When its requirements are met, it attains a height of twelve to fifteen feet. It is a large, coarse plant that crowds out cultivated plants in the garden or shades out other weeds in an open lot.

Though many of its leaves are trifid, that is, have three major lobes, many others on the plant have five lobes and the younger ones near the top of the plant are unlobed. Individual leaves on taller plants may attain a length of one foot. All are coarsely toothed and have stout petioles.

The staminate flowers shed pollen abundantly. The fruit that forms following pollination is about one-half inch or more long, brown in color, and has five or six ribs. A conical beak is found at the apex surrounded by five or six shorter spinelike structures that look like the points of a crown.

COMMON NAMES Wild hemp, giant ragweed, great ragweed, horseweed, bitterweed

SOME FACTS Native to the United States; annual; reproduces by seeds

RANGE Nova Scotia to Florida and west to northwestern United States, including Nebraska, Colorado, and Arkansas

HABITAT Moist, rich soil; fields; waste places

SEASON July–September

Ambrosia artemisiifolia
▶ **Hayfever weed, ragweed**

Hogs and sheep love the common ragweed, the same weed that causes so much human suffering. This accounts for one of its common names, hogweed. Another name, hayfever weed, is the more appropriate name for this plant.

The twice pinnatifid, parted two- to five-inch leaves look almost feathery on this one- to five-foot-high cousin of the taller *A. trifida*, the giant ragweed (p. 88). Its leaves are deep green above and much lighter green below, and are alternately distributed on a stem. Some of the leaves near the base of the stem are oppositely arranged.

The unattractive flowers are either male (staminate) or female (pistillate). Both are found on the same plant. The male flowers are displayed at the top on prominent spike-like racemes, while the pistillate flowers lower down are concealed behind clustered bracts. Hundreds of racemes of male flowers ensure a prodigious scattering of pollen and trouble for hayfever sufferers.

COMMON NAMES Hayfever weed, common ragweed, hogweed, bitterweed, wild tansy, roman wormwood, carrot weed, stammerwort

SOME FACTS Native to the United States; annual; reproduces by seeds

RANGE United States and Canada, from Nova Scotia to British Columbia south to Florida and Texas

HABITAT Dry soil, cultivated ground, waste places

SEASON July–September

Rumex
Genus

Rumex crispus
▶ Curly dock, yellow dock

Curly dock's wavy-edged, lance-shaped, longish leaves, borne near the top of the long taproot, are on longish petioles. Their bases are roundish or heart-shaped.

The one- to four-foot stem is smooth, ridged, and has swollen nodes. The inflorescences appear at the top of the stem, where branching is seen. The region of the inflorescences is usually branched several times but the branches are compacted, crowding together the drooping spikes of small six-sepaled greenish-yellow flowers.

Later in the season the stalk dies and turns a bright red-brown, each three-angled fruit maturing with the three inner sepals covering it. In winter *Rumex crispus* can be identified by its dead red-brown stalks topped by their branched inflorescences used frequently in dried flower arrangements.

Rumex and *Polygonum* are in the buckwheat family. The flour made from its seeds is comparable to flour made from buckwheat.

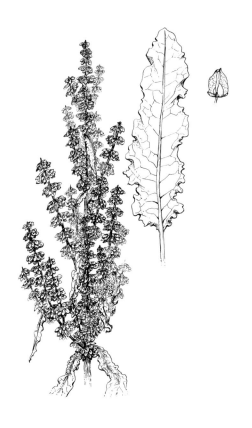

COMMON NAMES Curly dock, yellow dock, sourdock, narrow-leaved dock

SOME FACTS Introduced from Europe; perennial; reproduces by seeds

RANGE Throughout the United States and adjacent Canada

HABITAT Roadsides, waste grounds, fields

SEASON June–September

Matricaria
Genus

Matricaria matricarioides
▶ **Pineapple weed, mayweed, rayless chamomile**

Crush a bit of leaf material of pineapple weed and you will quickly learn why this epithet has been applied to this plant. Its pineapple-scented, finely-divided leaves are one to two inches long and may be found to be one to three times pinnatifid. Even without smelling, it may be immediately distinguished from *Anthemis cotula*, the stinking daisy, which it resembles, because pineapple weed lacks any sign of ray flowers, which the odoriferous daisy does have. Each floral head is a bluntly ovoid disc, greenish-yellow in color. Many are produced on each plant.

Its name, *Matricaria*, derives from its having once been found useful in treating infections of the uterus (mater, "mother," and caries, "decay").

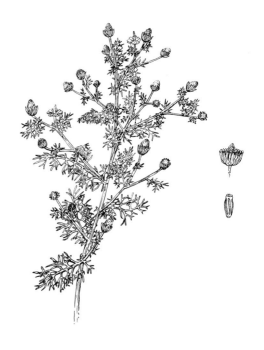

COMMON NAMES Pineapple weed, mayweed, rayless chamomile

SOME FACTS Annual; native to the Pacific slopes; reproduces by seeds

RANGE From Alaska to Baja California and east to Montana and Arizona; now in the Atlantic states, especially near cities and towns

HABITAT Roadsides, waste places, fields

SEASON May–September

Euphorbia
Genus

Euphorbia cyparissias
▶ **Cypress spurge, salver's grass, quacksalver's grass, graveyard weed**

Everyone knows the Christmas poinsettia (not poinsetta), *Euphorbia pulcherima*, with its red "flowers." The real flowers are quite small, yellow-green, and unstriking; it is the red leaves below the cluster of tiny flowers that produce the appearance of a huge colorful bloom.

The *Euphorbia* flowers are very reduced, having no calyx or corolla. A cup, called an involucre, which is top-shaped, surrounds several male flowers—each of which has but one stamen—and a single central female flower composed of a three-lobed pistil. Each cluster has greenish-yellow heart-shaped bracts below it. The many cups are produced in umbels.

The six-inch- to one-foot-long stems are erect, and may be many-branched. A milky, latex-containing sap is present in the stem. Simple linear leaves lacking petioles are distributed alternately along the stem. Each is about one millimeter wide.

The branching rootstock spreads cypress spurge far from the graveyard or garden where it was originally planted. Cypress spurge is, however, self-sterile, and thus if a colony springs up from but one plant by means of the spreading rootstock, seeds will not be produced.

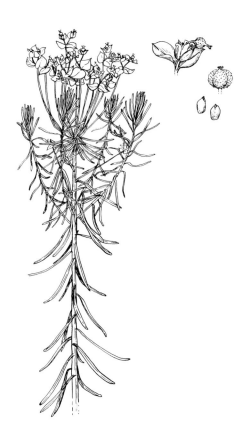

COMMON NAMES Cypress spurge, salver's grass, quacksalver's grass, graveyard weed

SOME FACTS Introduced as an ornamental from Europe; perennial; reproduces by seeds and creeping rootstocks

RANGE Widespread and locally very common in the northeastern states and north central states; occasionally on Pacific Coast

HABITAT Gardens, fields, waste places, roadsides, in older cemeteries

SEASON May–June

Portulaca
Genus

Portulaca oleracea
▶ Purslane, pursley

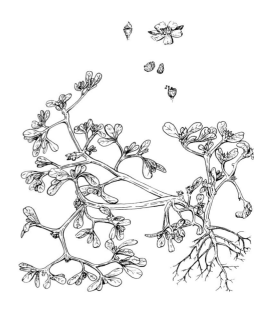

Purslane, also known as pursley, is a prostrate plant and forms many branches, giving rise to a mat. Its wedge-shaped leaves are simple and obovate. It also has thickened leaves filled with tissue for the storage of water, common in desert plants. Purslane is really a desert plant, though it has invaded other places. It uses the water-storage tissue in droughts.

The succulence is associated with a mucilaginous quality, which makes the leaves of this plant valuable as a thickener when added to soups and stews. They can be eaten cooked or raw and make a good potherb, though purslane has never attained importance as a potherb in the United States as it has in India, Middle East, and Europe.

From July to September small yellow sessile flowers open from flattened buds. The five yellow petals are inserted atop the sepals (which is uncommon among flowering plants) and open only in the sunshine. Following pollination, an urn-shaped, globular capsule results. It contains numerous seeds and opens by means of a round lid.

COMMON NAMES Purslane, pusley, pursley, wild portulaca

SOME FACTS Annual; reproduces by seeds; introduced from Europe

RANGE Widespread throughout Canada and the United States

HABITAT Gardens, cultivated fields, waste places; in rich soil and dry soil

SEASON July–September

Family

Lysimachia
Genus

Lysimachia nummularia
▸ **Moneywort, creeping jenny, creeping charlie**

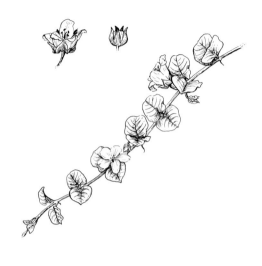

Moneywort's orbicular entire leaves look like coins and this feature has given the plant its name nummularia (from *nummulus*, Latin for "coin"). Several of its common names originate in the shape of the leaves.

The smooth stem the round leaves are attached to in opposite pairs creeps along the ground, rooting when it gets the opportunity and giving it the common names creeping loosestrife and creeping jenny. Since it grows rapidly and branches frequently, creeping jenny can cover a good-sized area with speed, forming a matlike growth.

One-inch-wide wheel-shaped yellow flowers are found in the axils of the leaves. The five-parted corolla of yellow myrtle is spotted with small dark red dots. This prostrate plant prefers damp ground, but can be found on a dry lawn or rock garden.

COMMON NAMES Moneywort, creeping loosestrife, herb twopence, two-penny grass, creeping jenny, creeping charlie, yellow myrtle

SOME FACTS Introduced from Europe as an ornamental; perennial; reproduces from seeds and creeping stems

RANGE Widespread in the northeastern states

HABITAT Fields, gardens, lawns, along ditches

SEASON June–July

Family

Hieracium

Genus

Hieracium pratense
▸ **Field hawkweed, king devil,
yellow devil, yellow paintbrush**

Field hawkweed is a fairly common weed in
fields, and, unfortunately, lawns. Except for
the yellow heads it produces atop a slender,
one- to two-foot-tall bristly stalk, it is very
much like its close relative, devil's paintbrush
or orange hawkweed (*Hieracium
aurantiacum*), which has bright orange-red
flowers. Both are formidable weeds.

   The flower-tipped scapose stem—there
are but two or three reduced leaves on it—
arises from a basal rosette of narrowly oblong
to lance-shaped leaves that taper back to
margin-bearing petioles. They are bristly-hairy
on both surfaces. Small runners may be
found on some plants.

COMMON NAMES Field hawkweed, king
devil, yellow devil, yellow paintbrush

SOME FACTS Introduced from Europe;
perennial; reproduces by seeds and stolons

RANGE Found more abundant in the
northeast states; less abundant west

HABITAT Fields, meadows, roadsides,
lawns, waste places

SEASON June–August

Family

Taraxacum
Genus

Taraxacum officinale
▶ **Dandelion, lion's tooth**

Lion's tooth (dents-de-lion) has sharp-pointed bright yellow "petals," which may, indeed, be very much like real lion's teeth. Each "flower" of the dandelion is actually composed of many individual flowers, and each "petal" is actually composed of five fused petals; so each "petal" indicates one entire flower. *Taraxacum officinale* is a member of the large taxonomic group known as the composites and is closely related to chicory (p. 188). Each flower is a ray flower.

Pick a "flower" of dandelion and pull it apart, spreading the pieces in the palm of one hand. Study each "piece" carefully—you will see that each is an entire flower!

The yellow flowering heads are produced from early spring to late, late fall (even into winter if the weather is warm). Each head, following seed set, becomes a ball of parachuted seeds; hence the common name blowball. These balls of easily detached parachutes have given pleasure to children for countless generations. Each bloom produces fifty or more parachutes, and each plant continues to produce heads of flowers throughout the growing season. Let one into your lawn and you are in trouble.

If you break the flowering stalk, you will note that it is hollow and that a milky substance appears at the site of the break. The milky juice contains latex, from which rubber could be made, though this is not commercially feasible at the present time.

COMMON NAMES Dandelion, lion's-tooth, blowball, cankerwort, milk witch, monk's-head, Irish daisy, priest's-crown, wet-the-bed

SOME FACTS Introduced from Europe, originally from Asia; perennial; propagates by seeds and by forming shoots from the taproots

RANGE Cosmopolitan weed

HABITAT Fields, waste places, meadows, lawns

SEASON March–December; will bloom during warm winters

104 Yellow-Flowered Species *Family*

Linaria

Genus

Linaria vulgaris
▸ Wild snapdragon, Jacob's ladder, yellow toadflax

Linaria vulgaris started out in America as a well-loved garden plant. It is a relative of *Verbascum*, foxglove, and is a member of the figwort family.

The racemes of inch-long bright yellow flowers with orange throats are produced near the top of the stem. Each flower is composed of five sepals and five petals fused into a two-lipped yellow corolla with a long orange throat. These bilaterally symmetrical flowers are pollinated by bumblebees. Moths also get the nectar, but by stealing. Instead of landing on the lower lip, depressing it, and getting doused with pollen, a moth will insert its long proboscis through the lips and take nectar without becoming involved in the pollination of the flowers.

The wild snapdragon's stems are erect and the plant is found in colonies; individual stems rarely branch. Simple, linear leaves are present on the stems, attached without petioles. They are pale green (almost blue-green) and seem pointed at both ends.

COMMON NAMES Butter-and-eggs, eggs-and-bacon, impudent lawyer, yellow toad-flax, ramsted, flaxweed, wild snapdragon, Jacob's ladder

SOME FACTS Introduced from Europe; perennial; reproduces by seeds and root-stocks

RANGE Eastern North America

HABITAT Cultivated land, meadows and pastures, roadsides (often on gravelly or sandy soil)

SEASON May–September

Oenothera

Oenothera biennis
▶ Evening primrose, field primrose, wild beet

The evening primrose, also known as field primrose and wild beet, first sends up from the top of its long, thick taproot a rosette of lance-shaped hirsute leaves some three to six inches long. During the second year a stem two to six feet in height, somewhat woody, emerges from the center of the rosette, and may branch from the base. Alternate leaves, which are narrow, lance-shaped, usually entire, and with short petioles, are found on the stem.

Bright yellow flowers that open are in the axils of the smaller leaves near the top of the stem. These sulphur yellow blooms are approximately one and a half inches in diameter and are generally closed during the day; however, flowers on individual plants are open during the day. As the blossoms open, a fragrance becomes apparent. As the sun rises, the blossoms droop.

After fertilization the ovaries grow into inch-long seed-filled capsules remaining on the old stems after they have dehisced at their summits. The seeds are very long lived.

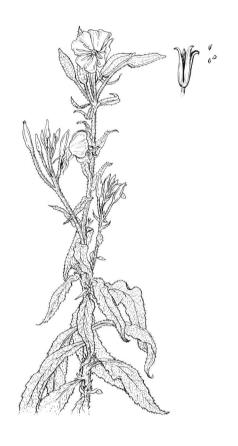

COMMON NAMES Evening primrose, field primrose, tree primrose, fever plant, night willowherb, wild beet

SOME FACTS Native (but now introduced to Europe); biennial; seeds

RANGE Labrador to Florida and west to Rocky Mountains

HABITAT Dry soil, fields, roadsides, waste places

SEASON June–September

Solidago
Genus

Solidago canadensis
▶ Canada goldenrod, tall yellowweed

Goldenrod is a very abused plant and is hated by hayfever sufferers even though they have no just reason for hating goldenrod, which merely produces golden blossoms at the moment hayfever sufferers happen to begin suffering. The two events are not related. Alabama, Kentucky, and Nebraska have taken this flower as their state flower.

Under excellent growing conditions, specimens of Canada goldenrod can grow three to six feet tall, even up to eight feet. It has alternate leaves narrowly lanceolate and essentially uniform from their base to their summit. Each is serrate, three-nerved, smooth along the upper surface, and hairy beneath. Leaves found lower down on the stem are petiolate, while those found higher up are usually sessile and entire.

The heads of this composite are borne along the upper surface of several flowering branches that curve outward and often downward. The individual heads are small and dull yellow, each with ten to twenty ray florets.

COMMON NAMES Canada goldenrod, tall yellowweed, tall goldenrod

SOME FACTS Native to North America; perennial; propagates by seeds and rootstocks

RANGE Widespread throughout eastern North America, occurs less frequently westward to British Columbia

HABITAT Rich, open soil; thickets, meadows, waste places

SEASON July–October

Solidago sempervirens
▶ **Seaside goldenrod**

The seaside goldenrod will probably never bother you as a weed unless you are fortunate enough to own some shoreline. Even then, this fundamentally attractive shore plant will very likely not disturb you. It lives in wet, usually salty places near the coast, and often forms a border approximately where high tide makes a line along the beach or shore.

The taxonomy of the genus *Solidago* is difficult, and even the experts may differ as to which species is which in some cases.

It has thick, succulent leaves, which are often noted on seaside or salt-loving species. The basal leaves are long-stalked. Like *Solidago canadensis* (p. 110), its heads of flowers are along the upper side of branches that curve outward.

COMMON NAME Seaside goldenrod

SOME FACTS Native to the United States; perennial; reproduces by seeds and rootstocks

RANGE Along coast from Gulf of St. Lawrence to Florida and Texas

HABITAT Saline places, brackish water, rear of beaches

SEASON August–October

Verbascum

Verbascum thapsus
Mullein, velvet leaf, velvet dock, great mullein

Velvet dock, felt wort, blanketleaf, flannel leaf—all refer to the hairs covering the leaves on both the rosette and the stalk.

A rosette of fuzzy leaves ten to twenty inches long and four to six inches wide appears during the first year of this biennial's growth. The leaves, because of the downy coat, appear white to gray-green.

During the second year's growth a stalk two to seven feet high grows from the center of the rosette, and may branch from one to two times near its top. The leaves that are found on the flowering stalk become progressively smaller from the bottom to the top of the flowering stalk. All of them, however, are densely hairy.

Yellow flowers open all along the stalk near the top and are nearly sessile, with very woolly sepals and sulphur yellow petals. The stamen hairs give one pollinator, the hover fly, a good grip as she lands. A globular capsule develops one-fourth inch long, each of whose two chambers contains many tiny brown seeds.

COMMON NAMES Mullein, flannel leaf, velvet leaf, velvet dock, hedge taper, candlewicks, blanketleaf, Aaron's rod, great mullein

SOME FACTS Introduced from Eurasia; biennial; reproduces by seeds

RANGE Abundantly found in north temperate America

HABITAT Field, roadsides, waste places, pastures; likes dry, stony, or gravelly soils

SEASON July–September

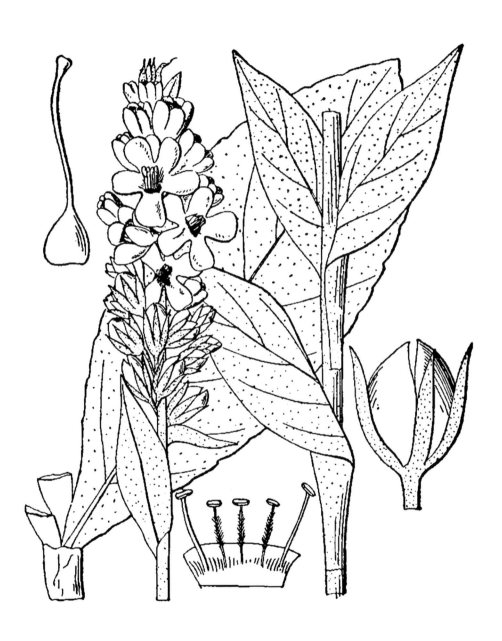

Verbascum
Genus

Verbascum blattaria
▶ **Moth mullein**

Some believe this weedy plant is called moth
mullein because night-flying moths pollinate
the blooms. Others assume it's because the
open flower looks like a moth.

Moth mullein's two- to five-foot stem has
few glandular hairs near the top and begins at
the center of the rosette and rises round,
slender, and usually unbranched. The petiolate lower leaves are oblong, though sometimes they may be pinnatifid, hairless, veiny,
dark green, and anywhere from three inches
to one foot long. Leaves higher on the stem
are sessile and pointed.

Both yellow- and white-flowered specimens are known, though the yellow is more
common. The upper petal is usually found to
be brown on its back, while the filaments of
the unequally sized anthers are coated with
purple hairs. Flowering begins at the bottom
of the open raceme with three-or four-inch-
wide blossoms appearing at one time. The
upper flowers are still opening after the lower
blossoms have produced their globose,
many-seeded capsules.

COMMON NAME Moth mullein

SOME FACTS Introduced from Europe;
biennial; reproduces by seeds

RANGE Quebec to Minnesota and south
to Florida and Kansas, west to the Pacific
and south to the Gulf of Mexico

HABITAT Meadows, pastures, waste places

BLOOMS June–November

Family

Helianthus

Helianthus annuus
▶ Sunflower

The sunflower's noncultivated form stands one to six feet high and the cultivated forms are up to fifteen feet tall. Broad, oval, pointed, cordate, three-ribbed leaves that are three inches to a foot long emerge from the tall, rough stem on stout, hairy petioles. Flower heads are three to six inches wide, with yellow, sterile ray florets, and the disc florets are five-lobed, tubular, dark brown, and fertile. Its well-known seeds (achenes) are ovate to sedge-shaped, flat, and white, gray, or dark brown with lighter stripes, or are gray-mottled.

The American Indians used the seeds for food and oil. They were parched and ground into flour, made into bread or cakes, or put into soup. The shells, roasted or crushed, are said to make a drink that approximates coffee in taste. It has been reported that the Indians of French Canada mixed the seeds into sagamite or corn soup.

Use your garden gloves when uprooting this very American plant because it is rough to the touch.

COMMON NAMES Sunflower, wild sunflower

SOME FACTS Native to the United States; annual; reproduces by seeds

RANGE Frequently seen east of the midwestern states, but native to region between Minnesota and Saskatchewan, and south to Missouri and Texas

HABITAT Meadows, fence rows, roadsides, waste places

SEASON July–September

Barbarea vulgaris
▶ **Winter cress, yellow rocket, St. Barbar's cress**

Nothing about winter cress can be directly compared to St. Barbara, protectoress of those caught in thunderstorms, besides the fact that the young leaf rosettes are available on December 4, St. Barbara's Day.

The green-in-winter rosette leaves, which are above the fibrous root system, and the leaves lower on the flowering stalk are pinnatifid, the terminal lobe being larger than the others. Upper-stem leaves are coarsely toothed and clasping. All the leaves of herb barbara are dark green, smooth, and shiny.

Branches of the one- to two-foot-long stem end in racemes of bright yellow four-petalled flowers, which form the Maltese cross pattern.

The leaves are used in salads, before late March. Flower buds may be cooked like cauliflower. The vitamin C-containing leaves are favored by cattle and sheep.

Inch-long pods (siliques) are seen not long after fertilization of the flowers. These are nearly erect and almost four-angled.

COMMON NAMES Yellow rocket, winter cress, St. Barbar's cress, bitter cress, herb barbara, rocket cress, water mustard, potherb

SOME FACTS Introduced to the United States a second time by colonists, who found it already here; weakly perennial (really biennial); reproduces by seeds

RANGE Abundant in northeastern and north central states, occasionally in Pacific Northwest

HABITAT Fields, meadows, roadsides, waste places

SEASON May–June

Ranunculus

Ranunculus acris
► **Tall field buttercup, tall crowfoot, butter flower, meadow buttercup, blister plant, goldcup, kingcup**

Thinking of the buttercup as a weed may seem strange to some, but in a poorly drained meadow or sections of some lawns, tall crowfoot can become a pest. Yet, we must admit that goldcup, weedy or not, is a charming plant. However, let your admiration remain ocular because blister flower produces an acrid (hence *acris*) juice that blisters the mouths and intestines of cattle and will blister your skin.

The inner surface of the cup of goldcup seems to glisten in sunlight. This is due to the presence of many starch grains in the cells of the epidermal (surface) layer of the petals.

COMMON NAMES Tall field buttercup, tall crowfoot, butter flower, meadow buttercup, blister plant, goldcup, kingcup

SOME FACTS Introduced from Europe; perennial; reproduces by seeds

RANGE Throughout the United States and Canada; northwest Washington

HABITAT Pastures, meadows, roadsides, waste places

SEASON May–September

Lotus
Genus

Lotus corniculatus
▶ Bird's-foot trefoil, hop o'my thumb

Bird's-foot trefoil, sporting its dark green foliage and bright yellow, typically papilionaceous blossoms, is lovely, but its spreading habit of growth causes it to grab space in which other plants might grow.

Hop o'my thumb, a winter hardy plant, has a stem that may be spreading, but if a number of plants are growing closely packed, the stems may be ascending or even erect. Under good growth conditions, they may reach a length of twenty inches. Sessile, alternate compound leaves are borne on the stems and each is composed of five linear to oval leaflets, the two lower of which are right next to the stem and resemble stipules. The bright yellow flowers occur in umbel-like stalked clusters in the axils of the upper leaves. Interestingly, the flowers may be found in shades from bright yellow to brick red. Following pollination, the pods cluster at the tip of the pedicel, and when mature, look like a bird's foot, hence the common name.

COMMON NAMES Bird's-foot trefoil, cat's-clover, crow-toes, hop o'my thumb

SOME FACTS Introduced from Europe; perennial; reproduces by seeds

RANGE Widespread in the United States

HABITAT Waste places, lawns, roadsides

SEASON June–August

Bidens
Genus

Bidens frondosa
▶ **Beggar ticks, stick-tights**

Beggar ticks become stuck to animals' fur and human clothing in an attempt to spread their seeds. Bidens are partial to poorly drained places, have freely branched shoots two to five feet high, and take on a purplish hue.

The three to five-parted compound leaves are oppositely disposed along the stems; a short distance below the terminal blossoms are found simple, lanceolate, toothed leaves, but a bit further down are three-parted compound leaves, and still further down, five-parted compound leaves. Each leaflet is lanceolate and sharply toothed. The central leaflet of compound leaves tends to be on a longer petiole than leaflets on the sides. All are smooth.

Beggar ticks is a composite, whose head of flowers is composed of two yellow-orange-flowers: ray flowers that are sometimes missing and disc flowers, which produce the seed. The fruit is a flat, black, wedge-shaped achene, ridged down the center. Two diverging points are at the apex; each is covered with downward-pointing barbs.

COMMON NAMES Beggar ticks, stick-tights, devil's bootjack, bur-marigold, pitch-fork-weed

SOME FACTS Native to the United States; annual; propagates by seeds

RANGE Widespread throughout the United States and into southern Canada

HABITAT Moist soil, especially in gardens, roadsides, pastures, waste places

SEASON July–October

Pastinaca

Genus

Pastinaca sativa
▶ **Wild parsnip, madnip, field parsnip**

A carrot relative, wild parsnip is also a biennial. During the first year there arise from the long taproot long pinnately-compound leaves that are composed of coarsely toothed and lobed segments, attached to the top of the taproot by long, flattened, and grooved petioles. The second season of growth there arises from the center of the rosette a two-to four-foot-tall hollow and deeply grooved light green small-leaved stem. Flat-topped compound umbels composed of yellow blossoms cap the stalk.

Madnip grows deep in the soil and is composed of phloem tissue filled with starch. It develops a sweet taste after exposure to cold.

American field parsnip is said to be poisonous during its second year; its European counterpart is thought not to be. Some people are sensitive even to the touch of the leaves and develop a rash.

Wild parsnip serves as host to a fungus that attacks the related celery plant, so it is necessary to remove it from fields of celery. As a weed it will be found under moist conditions.

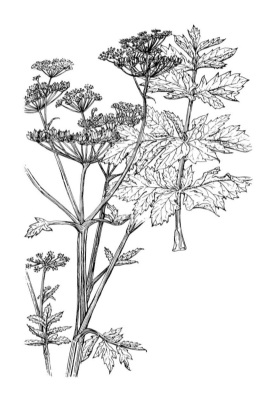

COMMON NAMES Wild parsnip, field parsnip, madnip, bird's-nest, hart's-eye, tank, siser

SOME FACTS Introduced to United States from Europe for its roots and escaped early; biennial; propagates by seed

RANGE Widespread in the United States and Canada

HABITAT Waste places, roadsides, meadows; on rich soil and usually where soil is moist

SEASON June–August

Tanacetum vulgare
▶ **Tansy, bitter buttons, ginger plant, parsley fern, hindhead, bitter weed**

The stem of tansy, also known as bitter buttons, ginger plant, and parsley fern stands one to three feet high and generally remains unbranched. It is the flowering part of the plant, at the top, that branches many times. Tansy's delicate leaves are beautiful. Each is deep green, smooth, and one- to three-pinnately divided. Leaf segments are narrow and toothed. Leaves higher on the stem are smaller and less divided than those lower down.

The common name bitter buttons refers to the button-shaped, clustered flowering heads (tansy is a composite). Each button is one-half inch across, yellow, and composed entirely of disc flowers. The central flowers are perfect, while the marginal ones are usually pistillate.

In England tansy is given the generic name *Chrysanthemum*, which is probably more accurate.

COMMON NAMES Tansy, bitter buttons, ginger plant, parsley fern, hindhead, bitter weed

SOME FACTS Introduced by colonists; perennial; reproduces by seeds and rootstocks

RANGE Widespread in the United States, especially in the Northeast; less frequently seen in the West

HABITAT Roadsides, waste places; old yards on dry, sandy, or gravelly soils

SEASON July–September

Potentilla

Genus

Potentilla canadensis
▶ **Common cinquefoil, five-finger, bareen strawberry**

The five-finger does not have five leaves, but three leaflets whose lateral two are so strongly parted they look like two separate leaves.

Common cinquefoil has very slender, hairy stems that are spreading and procumbent and may be two feet long. Stem tips touching the earth create new plantlets. The five-petalled flowers, yellow and one-half inch wide, are within the leaf axils of the first two nodes. Many stamens and pistils may be found around a plump central receptacle.

Do not confuse this species with the similar in appearance "silvery" *P. argentea*, also very common. Nor could you confuse the common cinquefoil with the strikingly robust and woody sulphur cinquefoil, *P. recta L.*, an introduced species which is perennial, erect, and may grow two feet tall.

COMMON NAMES Common cinquefoil, five-finger, bareen strawberry

SOME FACTS Native to the United States; perennial; propagates by seeds and stolons (runners)

RANGE Common from New England to the Great Lakes and southward

HABITAT Dry soil; fields, meadows, pastures, waste places

SEASON June–September

Family

Rudbeckia

Rudbeckia hirta
▸ Black-eyed Susan

It is difficult to dislike the lovely black-eyed
Susan, also called ox-eye daisy and coneflower. Its generic name is modeled after
Linnaeus' great teacher, Dr. Rudbeck, but
the truth is that once it invades a field, it
spreads very rapidly. Ox-eye daisy hails from
the plains and prairies of the West and
Midwest; no doubt it was a stowaway on
trains coming East.

During its first year a rosette of simple,
two- to six-inch-long, spatulate, three-nerved
leaves forms. These leaves are rough, hairy,
and have grooved petioles. An erect stem,
one to three feet high, which is simple or
sparingly branched near its base, arises from
the center of the first year's rosette. The stem
and the simple, oblong, sessile stem leaves are
also rough and hairy. Two- to four-inch-wide
heads occur singly. A rounded brownish-
black cone of disc florets is seen at the center
of the flower and is surrounded by fourteen
to sixteen long, sterile ray flowers that are
bright orange-yellow in color.

COMMON NAMES Black-eyed Susan,
brown-eyed Susan, coneflower, ox-eye
daisy, yellow daisy

SOME FACTS Native to the U.S.;
biennial; reproduces by seeds

RANGE Common in East, often seen in
Midwest and West

HABITAT Prairies, meadows, pastures,
waste places, old fields

SEASON July–October

Impatiens

Genus

Impatiens biflora

▶ **Jewelweed, touch-me-not, snapweed, wild lady's slipper, silver
cap, wild balsam, lady's eardrop**

Touch-me-not's inch-long seed pods (capsules) live up to this common name when
mature. The faintest disturbance and they
explode into five shreds, hurling their seeds
far and wide and even making a small but not
inaudible noise.

Jewelweed's glabrous (smooth) branching
stems are almost clear, something that may
be difficult to observe in the deep shade of
the stream banks where it lives and grows to
a height of two to four feet. Young shoots
may be cooked and eaten, though some say
touch-me-not is poisonous; however,
Professor John Kingsbury of Cornell
University does not list it among the poisonous plants in his book *Poisonous Plants of the
United States*.

The unwettable, thin, one- to three-inch
leaves of touch-me-not are pale green and
ovate to elliptical in shape. They are blunt-
ended, with crenately toothed margins, and
are attached on longish petioles.

The lovely orange-yellow flowers occurring in twos from slender pendent stalks are
mottled reddish-brown, and are spurred, but
the spur is on a petal-like sepal that ends in a
sac contracting into a long in-curved spur.

The stems are an antidote to poison ivy
when rubbed against an area recently in con-
tact with *Rhus radicans*. It prevents the
appearance of the blisters.

COMMON NAMES Jewelweed, touch-me-
not, snapweed, wild lady's slipper, silver
cap, wild balsam, lady's eardrop

SOME FACTS Native; annual; reproduces
by seeds

RANGE Newfoundland to South Carolina,
Arkansas, Alabama, and Oklahoma

HABITAT Moist woods, brooksides,
near springs

SEASON June–September

Family

Cuscuta

Cuscuta gronovii
▶ Dodder, gold-thread vine, onion dodder

Cuscuta, as well as many other angiosperm parasites, effects a connection with the host plant by means of its roots, which dig deep into the tissues of the stems of the host, stealing water, minerals, and foods. Though dodder has been shown to have a small amount of chlorophyll, and is, therefore, capable of some photosynthesis, it does not carry on enough photosynthesis to permit manufacture of all the food it requires.

Dodder will attach itself to just about any flowering plant. After germination from a seed in the soil, the stem, on reaching a neighboring plant of a different species, will send its roots into the stem tissues of the host. The longest single stem was six feet nine inches.

Dodder's stem has small scales rather than leaves. The yellow stem is due to the abundance of yellowish and orange carotenoids present. Dense clusters of small, waxen white flowers precede the production of globose capsules filled with rough or sticky seeds.

COMMON NAMES Dodder, gold-thread vine, onion dodder

SOME FACTS Native; reproduces by seeds and small pieces of overwintering stem; fundamentally an annual

RANGE Widespread in northeastern United States

HABITAT Low wet fields, marshy thickets, waste places

SEASON July–September

Polygonum
Genus

Polygonum aviculare
▶ **Knotgrass, matgrass, doorweed, pinkweed,**

Whole large sections of many lawns are composed of knotgrass (also known as birdgrass, stonegrass, and goosegrass), which indicates hard soil. When matgrass has been recently mowed, it gives the appearance of a blue-green grass. Indeed, growing pure stands of knotgrass might be one solution to the suburban lawn grower's problems.

Knotgrass leaves are oblong and narrow and can be seen in their bright blue-green color growing happily between the paving stones of the streets of large cities. They are one-quarter to one inch in length. The slender pale green stems are usually prostrate and measure four inches to two feet in length. Branches emerge from the joints (or "knots"), which appear paler within the sheath common to the genus *Polygonum*.

The tiny white to pink blossoms make their appearance either solitarily or growing in groups of two or three within the axils of the leaves. The whole perianth of a single flower is less than two millimeters long!

COMMON NAMES Knotgrass, matgrass, doorweed, pinkweed, birdgrass, stonegrass, waygrass, goosegrass, centinode, nine-joints, wine's grass

SOME FACTS Native to the United States and introduced from Eurasia; reproduces annually by seeds

RANGE Common in northern U.S. and southern Canada

HABITAT Hard-trampled lawns, yards, roadsides, paths, waste places

BLOOMS June–October

Family

Mollugo

Mollugo verticillata
▸ Carpetweed, Indian chickweed, whorled chickweed, devil's grip

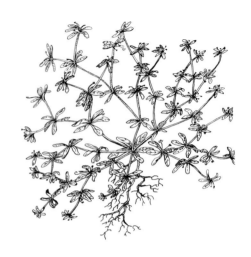

While there is not too much that can be said in favor of carpetweed, it can, in an emergency, be eaten, though you'd have to collect a very great deal of this small plant to make a stomach-filling repast!

Its stems are much branched, prostrate, and form mats. Simple, whorled leaves that are entire and spatulate appear in groups of five or six at each node. Small umbel-like flowers are found at the nodes on slender pedicels and in the axils of the leaves (verticillate). These flowers have five sepals, which are white inside and which have taken the place of the petals.

COMMON NAMES Carpetweed, Indian chickweed, whorled chickweed, devil's grip

SOME FACTS From tropical America but native to Africa; annual; reproduces by seeds

RANGE Eastern and middle United States to Florida and Texas

HABITAT Gardens, lawns, waste places, between the cracks of walks; dry, gravelly, and sandy soil

SEASON June–September

Stellaria

Stellaria media
▶ Common chickweed

Chickweed, also called starweed and bird-weed, is one of, if not the most common weed in America. Wherever the stems touch the ground the nodes give rise to roots and new stems. The plant continues to make new seeds through a mild winter, or gets a very early start in the spring. That makes the winterweed, as it is also called, a most successful plant.

The fundamentally five-parted flowers appear in the leaf axils or in cymose clusters, their white cleft petals being shorter than their green sepals. A small one-chambered capsule is produced in which tiny, approximately circular seeds are found, each measuring no more than one millimeter in diameter.

The low, slender branches of starweed are much branched and covered with rows of hairs.

Opposite leaves are produced, and they are ovate to oblong in shape. The higher leaves are largely sessile, while those lower down on the stem have hairy petioles.

COMMON NAMES Common chickweed, starwort, starweed, winterweed, birdweed, satinflower, tongue grass

SOME FACTS Introduced from Europe; annual; propagates by seeds

RANGE Worldwide; very common in North America

HABITAT Gardens, lawns, waste places, meadows, cultivated fields

SEASON Throughout the year if winter is mild

Fragaria
Genus

Fragaria virginica
▸ Wild strawberry

Interestingly, the strawberry is not a berry at all. Botanically, a berry is a fruit, i.e., a ripened ovary. What is consumed with such pleasure when one eats strawberries is the receptacle on which sit the pistils. The receptacle is a modified portion of the stem that, in the case of the strawberry, turns sugary and moist when mature. The true fruits are the tiny dots that are embedded in the surface of the "berry." (If they are embedded, you are eating the fruit of *Fragaria virginica*, but if they are flat against the surface of the receptacle, you are consuming the fruit of *F. vesca*.) The real fruit of strawberry is called an achene.

The little white flowers are very similar to the yellow ones of *Potentilla*. Both genera are very closely related members of the rose order (*Rosales*). Clusters of as many as twelve flowers may be found on pedicels of approximately equal length. Trifoliate toothed leaves are borne on longish petioles.

COMMON NAMES Wild strawberry, strawberry

SOME FACTS Native; perennial; reproduces by seeds and runners

RANGE Common in northeastern United States

HABITAT From Labrador and Newfoundland to Georgia

SEASON April–July

Family

Trifolium
Genus

Trifolium repens
▶ **White clover, honeystalk, lamb suckling, husbandman's barometer**

A four leaf clover is actually four leaflets of one leaf and not four leaves. Each leaflet of the leaf is notched at the outer end (obcordate), and you will find that the leaflets fold at night or at the approach of a storm when the sky darkens. This helps account for the common name of husbandman's barometer.

The small flowers of white clover are pollinated by bumblebees, which take nectar (which is over 40 percent sugar) from the small, numerous flowers of the round head. After pollination, each flower bends downward and then turns brownish while the fruit and seed mature.

The stems root at every node, which accounts for the very rapid spread of white clover (it will very soon fill all the spaces of your bluegrass lawn).

It is interesting that the leaves have been found to be cyanogenetic, yet the species has been used for generations as cattle feed.

COMMON NAMES White clover, ditch white clover, white trefoil, purple grass, purplewort, honeystalk, lamb suckling, honeysuckle clover, trinity leaf, shamrock, husbandman's barometer

SOME FACTS Native of Europe; perennial; reproduces by seeds and has a creeping stem

RANGE Newfoundland to Alaska and south to Florida and California, but more abundant in north

HABITAT Fields, lawns, copses

SEASON May–September

Family

Silene

Genus

Silene cucubalus
▶ Bladder campion

The bladder campion's "bladders" are composed of thin-textured sepals, fused to form the urn-shaped bladder. This has about twenty observable pinkish interwoven veins. The small urn-shaped capsule has six teeth at its top and opens there as well.

Leaves that are opposite and sessile are found along the stem and each tapers to its tip. Each of the five petal edges is divided into two lobes, suggesting ten rather than five petals.

The young plants are palatable when cooked—they taste like peas, though they are somewhat bitter because of the presence of saponin, a chemical common in members of the pink family. Saponins lather in water, but in large amounts are dangerous because they dissolve the red blood cells. *Saponaria officinalis* (p. 178), our soapwort, has leaves with much saponin in them. Cooking destroys much of the saponin. In 1685, when most of the crops on the island of Minorca failed, the population fell to dining on bladder campion. It also makes a perfectly good fodder.

COMMON NAMES Bladder campion, bubble poppy, white bottle, cowbell, maiden's tear

SOME FACTS Perennial; reproduces by seeds and rootstocks; originally from north Africa; came to United States through Europe and was naturalized here

RANGE Common now in the East and widely distributed throughout temperate North America; local in Pacific Northwest

HABITAT Waste places, grasslands, roadsides, cultivated ground, meadows

SEASON June–September

Family

Lychnis

Genus

Lychnis alba
▸ White campion, white cockle, evening lychnis

White cockle's spreading, branching stem rises to two or three feet and is hairy and slightly sticky, particularly in its uppermost parts. Its simple, opposite leaves are ovate-lanceolate and also hairy. The lower leaves taper to margined petioles, while those higher on the stem are sessile.

The five-petalled flower may be staminate or pistillate; staminate flowers occur on separate plants from those bearing pistillate flowers. Both types of flower occur in loose cymose clusters, though solitary flowers are frequently seen in the field.

The five-fused sepals of the pistillate flower form an inflated calyx that is red-tinged along the hairy veins. The calyx is less swollen in the staminate blossom. All five white or pink petals of the fragrant inch-wide flowers are notched on their outer edges, which may falsely suggest more petals than are actually present. Five styles may be seen at the top of the floral tube. The capsule contains numerous seeds and splits open by ten teeth at its top.

COMMON NAMES White campion, white cockle, evening lychnis, snake cuckoo, thunder flower, bull rattle, white robin

SOME FACTS Introduced from Europe; perennial or biennial; reproduces by short rootstocks and seeds

RANGE Eastern United States and Canada; locally common in north central states and Pacific Northwest

HABITAT Waste places, lawns, grain fields

SEASON June–August

Family

Solanum
Genus

Solanum nigrum
▶ **Black nightshade, deadly nightshade, poisonberry, hound's berry**

Poisonberry's roughly rhombic, toothed leaves are alternately displayed upon a much branched stem one to two feet high. The leaves are not armed as they are in the closely related horse nettle (also called wild tomato), *Solanum carolinense*. In the latter species the leaves sport stout yellowish prickles along their veins, midrib, and petiole.

Deadly nightshade's small, white, relatively wheel-shaped flowers have five petals, which are produced in small, umbel-like drooping clusters at the side of the stem but near its top. (These blossoms are very similar to those of horse nettle, *S. carolinense*, but smaller.) The berries that form following fertilization turn from green to black. Deadly nightshade's fruits may be poisonous, since they contain solanine, a toxic glyco-alkaloid. Whether or not its berries are poisonous, deadly nightshade's leaves and stems definitely are not and may be safely used as a potherb with proper cooking.

COMMON NAMES Black nightshade, deadly nightshade, poisonberry, hound's berry, stubble berry, garden nightshade

SOME FACTS Native to the United States; annual; reproduces by seeds

RANGE Common all over the United States

HABITAT Fields and waste places; common on loamy or gravelly soils

SEASON July–September

Family

Apocynum

Genus

Apocynum cannabinum
▶ **Dogbane, American hemp**

Dogbane's one- to five-foot-long branching stem, whose main axis is exceeded by the branches, arises from a deep, vertical, branched root.

Oblong, opposite, two- to four-inch-long leaves emerge from the stems on short petioles or are sessile. If a leaf is removed, a milky juice will appear.

The terminal or axial flowers, in dense clusters that are flat-topped, are not strikingly colorful or large. Each blossom is five-lobed and greenish-white. The twin four-inch pods (follicles) that result from fertilization are round and smooth, and while maturing, relatively conjoined at their distal ends, though they can be easily separated by a touch. Each pod is filled with numerous brownish flat seeds attached to a tuft of fine white hairs. Thus parachuted, the seeds are carried far and wide by the wind.

From the much-branched root is obtained an extract containing several resins and glycosides, some of which have a stimulating effect on the heart.

COMMON NAMES Dogbane, American hemp, Indian hemp, Choctaw root, bowman root, dropsy root, blind hemp

SOME FACTS Native; reproduces by seeds and rootstocks; perennial

RANGE Widespread in the United States

HABITAT Fields, thickets, moist soils

SEASON June–August

Family

Chrysanthemum

Genus

Chrysanthemum leucanthemum

▶ **Ox-eye daisy, poorland flower, marguerite**

Ox-eye daisy, the state flower of North Carolina, is a well-known and well-liked plant but will take over a pasture with its spreading rootstocks, as well as the many seeds it makes.

The twenty to thirty white "petals" are not individual petals at all. Each petal is composed of five false petals fused together. If you study their edges closely, you will be able to see a waviness, which indicates the outline of the five. In addition to the white ray flower, there is the golden center, or "eye," composed of tubular disc flowers, which are each in turn composed of five hairlike sepals, five fused petals, stamens, and two fused pistils—a complete flower.

The head is terminal on a one- to three-foot-long stem. Several stems may be produced from each root crown, each having toothed, simple, alternate leaves. The leaves along the stem differ from the rosette at the base of the stem in that the higher leaves are sessile, while those of the rosette are petiolate and pinnately-lobed as well as oblong or lanceolate.

COMMON NAMES Ox-eye daisy, white daisy, white weed, field daisy, poorland flower, marguerite, poverty weed, moonpenny, dog blow, maudlin daisy

SOME FACTS Introduced from Europe; perennial; reproduces from seeds, and not very effectively from short rootstock

RANGE Common in northeastern United States; also found southward, and westward to the Pacific Coast

HABITAT Waste ground, oil fields, pastures, gardens

SEASON June–July

Phytolacca
Genus

Phytolacca decandra
▶ **Pokeberry, inkberry, poke,
red-ink plant**

Inkberry produces a root of large size that
contains phytolaccin, a cathartic. Pigs have
been poisoned from ingesting pieces of the
root of pokeberry, dying of bleeding gastritis.
Man can also be poisoned by pokeberry root.

Poke may grow to be ten feet tall. Older
stems turn purple late in the season. These
stout stems may branch above, and are covered with alternate, simple, entire leaves with
long petioles. Fundamentally of an oblong-
lance shape, some may reach the length of
one foot, and when bruised, emit a somewhat unpleasant odor.

Smallish white flowers are borne in
racemes; each flower has ten stamens, the
source of the specific epithet (deca, "ten";
andra, "male organ").

The American Indians used pokeroot
decoctions (as did the early American set-
tlers) for a variety of illnesses.

COMMON NAMES Pokeberry, inkberry,
poke, red-ink plant, scoke, pigeonberry,
garget, pocan bush, charges

SOME FACTS Native; European gardeners
have adopted it; perennial; reproduces by
seeds and rootstocks

RANGE Common in nearly all of temperate
North America

HABITAT Waste places, fence rows, thickets

SEASON June–September

Datura

Datura stramonium
▶ **Jimson weed, Jamestown weed, thorn apple**

Jimson weed is a large flowered plant rising to a height of five feet. Its pale green stems branch by forking; attached to them are alternate, large, oval but irregularly lobed and toothed ill-scented leaves. Attached by stout petioles, they are darker green above than below, and are large-veined.

Within the forking of the branches appear the solitary three- to four-inch-long trumpet-shaped flowers on a short peduncle. Each of the five white petals has a prominent tooth. A five-lobed and ridged calyx surrounds the petals, reaching halfway up the floral tube. Flowering is followed by the production of two-inch-wide oval, spiny capsules, which when dry open at the top. These contain four cells and many dark brown, flat, kidney-shaped, three-millimeter-long, and very poisonous seeds.

Every part of this plant is potentially lethal. Several powerful alkaloids are found in its tissues and all are poisonous when consumed in small quantities.

COMMON NAMES Jimson weed, Jamestown weed, thorn apple, mad apple, stinkwort, angel's trumpet, devil's trumpet

SOME FACTS The origin of this weed is in doubt—some think it originated in India, and others that it was introduced from the tropics; annual; reproduces by seeds

RANGE Florida to Texas, and north to Canada

HABITAT Fields, waste places, dumps

SEASON June–September

Eupatorium
Genus

Eupatorium perfoliatum
▶ **Boneset, thoroughwort, ague-weed, fever weed, sweating plant, crosswort**

The Doctrine of Signatures assumes that any plant organ that looked like a human part was capable of yielding a remedy for the sickened human part. Since the opposite, wrinkled, pointed leaves of boneset are united at their bases, the stem passing through the area of fusion (per, "through"; folia, "leaf"), it was obvious to our ancestors that a poultice made of the leaves and applied to a broken bone would help the parts rejoin. However, boneset is still used as a home remedy to dissipate a cold by inducing sweating.

The fused leaves are found on a two- to six-foot-high rather stout stem that branches at the top. Unlike its close relative, Joe-pye weed (p. 196), the compact corymbose clusters of small perfect flowers are a dull white, though occasionally one does find a boneset with blue flowers.

COMMON NAMES Boneset, thoroughwort, ague-weed, fever weed, sweating plant, crosswort

SOME FACTS Native to our shores; perennial; reproduces by seeds

RANGE Widespread in eastern North America, westward to Nebraska and Texas

HABITAT Wet fields, pastures, swamps, sides of streams and ditches

SEASON August–September

Lepidium

Genus

Lepidium virginicum
▶ **Peppergrass, bird's-pepper, poor man's pepper, tongue grass**

Lepidium sativum, a European relative of our native peppergrass, is a salad plant well known for its peppery but interesting taste. Though ours is peppery too, it is unfortunately not as tasty, except possibly to birds, which devour the small fruits by the thousands (thus giving the plant one of its common names, bird's-pepper).

The much-branched stem is six to twenty-one inches in height. The lower leaves are pinnatifid, approximately spoon-shaped in outline, their terminal lobes large and their lateral lobes very small. Leaves that are higher on the stem are lance-shaped and smooth.

The elongating racemes of white-greenish four-petalled flowers produce, after fertilization, two-celled pods (siliques), which are smaller and rounder than those of the plant's cousin, *Capsella*. These tiny pods are also flat and notched at their upper end, but unlike their many-seeded relative, they contain only two reddish-yellow seeds.

COMMON NAMES Peppergrass, bird's-pepper, poor man's pepper, tongue grass

SOME FACTS Native; annual; propagates by seeds

RANGE Newfoundland to South Dakota, southern Florida to Texas

HABITAT Waste places, roadsides, dry soils

SEASON May–September

Capsella
Genus

Capsella bursa-pastoris
▶ Shepherd's purse

Shepherd's purse, also known as pepper-plant, is one of the most common of the weeds in the U.S. and is related to mustard, the wallflower, cabbage, radish, and other valuable edible plants, and to many other garden plants.

A rosette of alternate, simple, toothed leaves appears first on the surface of the ground, while under the soil is a thin taproot. From the rosette's center arises a thin, erect, branched stem with leaves, but these leaves are unlike those of the rosette. They clasp the stem and are arrow-shaped because of the small pointed ears at their bases. At the tip of the branch a slender lengthening raceme of small, four-petalled white flowers, with the typical cruciform petal arrangement, is produced.

Flowers continue to open at the top of the raceme, while mature (triangular) fruits (siliques) are found lower down on the same stem.

At maturity the fruits open to shed several seeds, each about one millimeter in diameter. A single plant may produce two thousand or more seeds each year.

COMMON NAMES Shepherd's purse, pepperplant, caseweed, pickpurse

SOME FACTS Introduced from Europe; annual; reproduces by seed production

RANGE Common throughout the United States

HABITAT Gardens, cultivated fields, waste places

SEASON March to November in northern United States; all year elsewhere

Polygonum
Genus

Polygonum cuspidatum
▶ Japanese knotweed, Mexican bamboo

Polygonum cuspidatum, also called Mexican bamboo and Japanese knotweed, can take over a garden in very short order. The three-to nine-foot-tall stems do resemble bamboo stems, but are not true bamboo.

The underground stem or rhizome, which may attain a length of five or more feet, makes this persistent plant difficult to root out. In the fall the erect portion dies back to the rootstock, and this adds to the ugliness of *P. cuspidatum* in a garden.

The plant's leaves are broadly ovate, relatively truncated at their bases, and attached to the stem by petioles. Around the nodes of all members of the genus *Polygonum* are sheaths which may be described as papery, cufflike masses of tissue.

The apetalous flowers are small and greenish-white and are found in branched panicles. Though small, the sepals are showy floral organs and, interestingly for a dicot, there are three fused pistils with three separate stigmas. As the fruit forms, the three lower membranous sepals enclose it.

COMMON NAMES Japanese knotweed, Mexican bamboo

SOME FACTS Native of Japan; perennial; propagates by seeds and by rhizomes

RANGE Widely found in the northeastern part of the United States, west to Minnesota and Iowa, south to Maryland

HABITAT Waste places, neglected gardens

SEASON August–September

Daucus
Genus

Daucus carota
▶ Wild carrot, Queen Anne's lace, bird's-nest weed

Good Queen Anne's lace collars were delicate and lovely, as is the delicate compound inflorescence of wild carrot. Each tiny umbellet of the compound umbel is seen to be composed of small white flowers, and at the center blooms a single flower, almost black in its deep purpleness.

After pollination, during seed set, the outer umbellets bend inward, causing the whole inflorescence to appear concave and look much like a bird's nest. It is during seed maturation that an oil, smelling something like turpentine, can be extracted.

Though called wild carrot the plant will not produce the sweet, thick orange taproot.

A crown of lacy leaves is also produced during the first year's growth, each twice or three times pinnate, giving the rosette a feathery appearance. The following year the flowering stalk makes its appearance and bears relatively few sessile or clasping leaves. If crushed, the leaf gives off a characteristic and not too unpleasant odor.

COMMON NAMES Wild carrot, Queen Anne's lace, bird's-nest weed

SOME FACTS Introduced from Eurasia; biennial; propagates by seeds

RANGE Common throughout North America

HABITAT Waste places, meadows, pastures, roadsides

SEASON June–September

Achillea
Genus

Achillea millefolium
► Yarrow, milfoil, soldier's woundwart, thousandleaf

Achilles is said to have used soldier's wound-wort to staunch his soldiers' wounds at Troy. Yarrow has definite astringent properties, making the common name woundwort truly descriptive.

Thousandleaf stands one to three feet tall and is sometimes branched above. Each sessile leaf is bipinnately dissected into fine divisions, which cause the plant to appear to have a thousand leaves. Each leaf may reach a length of ten inches, though half that length is more usual. The leaves are deep green, and if one is chewed, a bitter aftertaste is left in the mouth. If the foliage is bruised, it is found to be strong-scented. Cattle avoid eating these leaves.

Its flowers are produced in dense, flat-topped compound corymbs. Individual heads of flowers are relatively small if compared with another composite, such as the daisy, and each head is composed of both ray and disc flowers. The small petals may be white or pinkish.

COMMON NAMES Yarrow, milfoil, knight's milfoil, thousandleaf, bloodwort, soldier's woundwort, nosebleed weed

SOME FACTS Native; perennial; propagates by seeds and by rootstocks

RANGE Throughout most of North America and in most parts of the world

HABITAT Waste places, meadows, pastures, roadsides

SEASON June–October

Family

Melilotus

Melilotus alba
▸ White sweet clover, white melilot, king's clover

White sweet clover is used as a forage plant, in which instance it is beneficial in two ways: it adds nitrates to the soil because it has nitrogen-fixing bacteria within the nodules on its roots, and the shoot is good for animals to eat.

King's clover grows to be three to eight feet high, while the yellow melilot usually reaches only two to four feet in height. Hart's clover has a smooth, slender, much-branched stem. The leaves of both species are pinnately trifoliate, and each leaflet is oblong to elliptical with very fine teeth. At the petiolar base are two stipules, both shorter than the leaf.

The typical leguminous flowers are produced on one side of a thin, spikelike raceme appearing in the axils of the upper leaves. The white quarter-inch-long blossoms are more fragrant than their yellow cousin's. Both produce small ovoid, wrinkled pods containing a few very long-lived seeds. Those of the white clover have a coarse network of ridges while those of the red clover are cross-ridged.

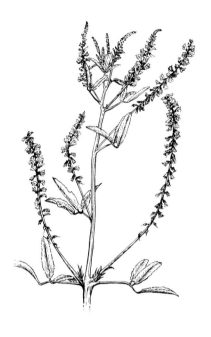

COMMON NAMES White sweet clover, white melilot, honey clover, tree clover, honeylotus, cabul clover, bokhara clover, king's clover, hart's clover

SOME FACTS Introduced from Eurasia; biennial; reproduces from seeds

RANGE Widespread throughout the United States

HABITAT Sandy or gravelly fields, roadsides, waste places

SEASON June–October

Saponaria

Saponaria officinalis
▶ **Soapwort, scourwort, wild phlox**

One underground rhizome will give rise to many stems, thus accounting for the great number of individual plants usually found growing together. Individual plants of bouncing bet may grow two or three feet high, and are smooth with swollen joints. The stems may remain unbranched or may branch sparingly.

Wild phlox has opposite, simple leaves that lack petioles and are ovate-lanceolate in shape with smooth margins. Close inspection reveals what appears to be parallel venation but is actually palmate venation.

The lovely pink five-petalled blossoms (the petals are fused) make their appearance in dense terminal clusters, though others may appear in the axils of opposite leaves lower down on the stem. The five sepals are seen to be fused into a long tube at the base of the flower.

Soapwort should be poisonous, as the soap-forming principals are poisonous, but few reports of poisoning have been verified. Animals generally seem to find the weed distasteful.

The plant has been known since the days of Dioscorides and has been used to treat venereal disease, though this use has long been abandoned. *Saponaria* causes trouble in the grain crop grown in the Northwest U.S.

COMMON NAMES Soapwort, bouncing bet, fuller's herb, scourwort, wood phlox, mockgilly flower, hedgepink, wild sweet william, wild phlox

SOME FACTS Introduced from Europe; perennial; reproduces by seeds and rhizomes

RANGE Temperate North America

HABITAT Along railroad embankments, roadsides, empty lots, waste places

SEASON July–October

Rumex

Rumex acetosella
▶ **Field sorrel, red sorrel, sourweed, sour leek, little vinegar**

Sourweed spreads rapidly on acidic ground, and where nitrates have been leached from the soil. Sorrel means "sour." Red sorrel's slightly bent stem never gets larger than a foot high. It produces rapidly spreading rhizomes.

The leaves of this weed are interesting in that they are halberd-shaped (hastate) and are spicy in their sourness when bitten.

The sourness arises from the presence of crystals of calcium oxalate, a dangerous poison that is deposited in the leaves; however, a few leaves of red sorrel will not harm anyone, and if used to make sorrel soup, the boiling will reduce the oxalate content drastically. The female flowers are tipped with three tiny bright red feathery stigmas, and the whole flower (female) is red at maturity (hence the common name red sorrel). When the stalks with their mature female flowers are all up, the field or roadside takes on a red glow that is not unattractive. The inflorescence takes up approximately half the stem of this relatively short plant.

In the eighteenth century, blue was dyed with indigo, but to get black, the leaves of the common field sorrel (*Rumex acetosella*) were boiled with the material to be dyed, which was then dried and boiled again with logwood and copperas.

COMMON NAMES Field sorrel, red sorrel, sourweed, sour leek, little vinegar plant, cow sorrel, horse sorrel, sour dock

SOME FACTS Introduced from Europe; perennial; reproduces by seeds and creeping rootstocks

RANGE Throughout the United States and Canada

HABITAT Old cut fields, pastures, meadows, highway embankments

SEASON May–October

Trifolium pratense
▶ Purple clover, red clover

The stems of red clover may become two feet long, and may either grow erect or rest on the ground. These stems are covered with soft hairs.

Purple clover's leaves are divided into the three leaflets expected in clovers and have V-shaped whitish areas across the middle of each leaflet, the point of the V aimed at the point of the leaf. Lower down on the stem the leaves are on petioles, while higher up they tend to be sessile.

The "flower" of clover is a cluster of many flowers, globose in shape, red or purple in color, and terminal upon the stems that bear them. Each individual flower of the cluster is typical of the legume-type flower, with a banner upper petal, wings, and a keel.

After pollination by bumblebees, each individual blossom is lowered from its former relatively upright position in the cluster and turns brown as seed set occurs.

Red clover is frequently planted for forage. Digestion of large quantities of red clover makes horses oversalivate (called "the slobbers") and has been known to impair or destroy the vision of cattle. In spite of this John Gerard noted in 1597 that "oxen and other cattell do feed on the herb as also calves and young lambs. The flours are acceptable to bees. Pliny writeth and setteth it downe for certaine that the leaves hereof do tremble and stand right up against the coming of a storme or tempest."

COMMON NAMES Purple clover, red clover

SOME FACTS Introduced from Europe; perennial; reproduces by seeds

RANGE Throughout temperate North America

HABITAT Lawns, meadows, pastures

SEASON May–August

Vinca
Genus

Vinca minor
▶ Periwinkle, myrtle, small periwinkle

The lovely blue funnel-form flowers of periwinkle make it difficult to think of this plant
as a weed, but it spreads so fast by means of
its runners that it can rapidly become weedy
and take over areas where it is not wanted.
Perhaps this is because it has forty-six chro-
mosomes (precisely the chromosome number of the human being). We, too, seem
rapidly to take over areas where we are not
entirely wanted.

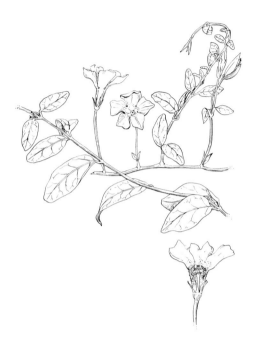

This relative of the dogbanes (p. 156) has
creeping stems that root at the nodes. Along
it are dispersed the simple entire oblong to
ovate leaves that are firm to the touch, glossy,
and evergreen. Its solitary blue flowers with
their five petals forming a funnel-shaped
corolla are within the axils of the leaves and
have truncated lobes on their edges.

While periwinkle produces fruits (two
short cylindrical follicles), reproduction by
seed is far, far rarer than by runners.

John Gerard in 1598 reminded his readers
that "the leaves boiled in wine and drunken,
stoppeth laske and bloodie flux; it likewise
stoppeth the inordinate course of the mon-
ethy sicknesse."

COMMON NAMES Periwinkle, myrtle,
small periwinkle

SOME FACTS Perennial; arrived from
Europe as a garden plant; propagates
through runners

RANGE Widespread in northeastern
United States; also on Pacific Coast

HABITAT Moist, rich soil; bordering lawns,
gardens, roadsides, cemeteries

SEASON February–June

Echium vulgare
▶ **Blueweed, viper's bugloss, blue devil, blue thistle, viper's herb, snake flower**

Blueweed is rooted to the soil by a sturdy, dark, long taproot. During the first of its two years of growth a rosette of entire oblong to lanceolate leaves up to six inches in length appears; all of the leaves are covered with bristly hairs on both surfaces and are dark green.

Blueweed grows on gravelly shaley slopes, and may dominate the landscape. During the second year of growth a bristle-covered stalk grows from the center of the rosette and may reach one to three feet in height. The gray-green bristles begin as stiff hairs and harden into outright prickles, arising in a red tubercule that specks the stem.

The lovely flowers of blue thistle are produced in curved racemes (sometimes called cymes), which arise in the upper axils. A five-lobed, irregular, funnel-shaped corolla is pink in the bud, and bright blue at maturity. Since the stamens remain red, they stand out against the bright blue of the mature blossom and the pollen they shed is among the smallest known of the flowering plants, only one to fourteen microns (1 micron = 1/1000 mm).

The open blossoms are not fragrant to us, but the bumblebees visiting them apparently are aware of a scent. After pollination, four small wrinkled nutlets result and are very long lived. They are frequently found as constituents of other batches of seeds.

COMMON NAMES Blueweed, viper's bugloss, blue devil, blue thistle, viper's herb, snake flower

SOME FACTS Introduced from Europe; biennial; propagates by seeds

RANGE New Brunswick to Ontario, south to Nebraska and Georgia

HABITAT Meadows, pastures, waste places, poorly drained slopes, roadsides

SEASON June–September

Cichorium

Cichorium intybus
▶ **Chicory, succory, ragged sailors, coffeeweed, blue daisy**

Cichorium intybus's leaves make a fine salad base or potherb, though one should choose the young leaves of the rosette. Succory is a first cousin of endive; its taproots are tasty cut up and placed in salads, or dried, powdered, and used as a coffee adulterant.

Silver-dollar-sized bright blue flowers are produced in clusters of one to four each on very short pedicels. Only one flower of each cluster blooms at a time—a very short time, opening early in the morning and closing by noon on bright days, but remaining open for a longer period on darker days. The blue daisy is not exactly similar to the real daisy, for the real one has both ray and disc flowers, while the blue version has only ray flowers (as do dandelions). Withering of the flower is independent of the pollination of the flower.

Ragged sailors' blossoms are distributed along a two- to five-foot hollow stem that is sparsely branched. The leaves are roundish, four to eight inches long, and alternately distributed along the stem; those lower on the stem are smaller than those of the rosette at the base of the stem, and are clasping and eared. Those of the rosette are spatulate, narrowing into marginal petioles. If you look at the underside of cornflower's rosette leaf, you will see stiff hairs along its midrib.

COMMON NAMES Chicory, cornflower, succory, ragged sailors, witloof, wild endive, coffeeweed, blue daisy

SOME FACTS Introduced from Europe; perennial; propagates by seeds

RANGE Nova Scotia to Florida, west to the plains

HABITAT Vacant lots, fields, roadsides

SEASON July–October

Cirsium

Cirsium arvense
▸ **Canada thistle, creeping thistle, green thistle**

Canada thistle or creeping thistle is a very difficult plant to get rid of once it has taken hold because it produces rapidly growing rhizomes, which send up above-ground shoots at short intervals. Almost every state in the Union has outlawed it!

The erect, grooved stem may be one to four feet high and is woody. Its heavily-armed leaves are three to six inches long, sessile, and irregularly pinnatifid. Each is toothed with hard, white needlelike spines that point this way and that.

Cirsium, the small-flowered thistle, is a composite, and therefore produces its flowers in heads: these are terminal or axial clusters, and pink or lavender-purple in color; all are disc flowers. The plants are single-sexed, which is a blessing. If only one sex has sprung up in your garden, your problem will be to fight the impossibly fast-growing rhizome; you will not have to worry about the seeds.

Mature seeds must wait until the plant dies and the heads are dropped to be freed, in spite of the fact that a rich supply of down is associated with the seeds in the head. When the head matures, the down flies away unattached to anything. Seedlings are rarely encountered.

COMMON NAMES Canada thistle, creeping thistle, green thistle

SOME FACTS Introduced from Eurasia; perennial; reproduces by rhizomes, rarely by seeds

RANGE Common to southern Canada and northern United States, as far south as Virginia and northern California

HABITAT Pastures, meadows, cultivated fields, waste places

SEASON July–October

Glechoma
Genus

Glechoma hederacea
▸ **Ground ivy, gill-over-the-ground, field balm, creeping charlie, cat's-foot, runaway-robin, ale-hoof, gillale**

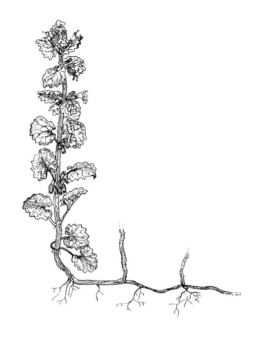

The appearance of the word "ale" in two of the common names given above gives a strong hint of one of the uses of this very common lawn weed. It was once used in place of hops for the flavoring and preparing of homemade ale. The gill in the common names points to the same use, for gill comes from *giller*, which is French for "ferment."

On its creeping, sometimes eighteen-inch-long stem, ground ivy sports small orbicular leaves with crenate edges. That this plant is a member of the mint family is given away by two things: its square stems, and its two-lipped flowers, which are bilaterally symmetrical.

Clusters of these pale purple flowers appear in the axis of the leaves and give rise to nutlets that are brown but have one white spot.

The late Euell Gibbons reminded us in *Stalking the Healthful Herb* that this plant can be used as a bitter tonic and nutritious tea that is an excellent remedy for a stubborn cough, and that it is loaded with vitamin C.

COMMON NAMES Ground ivy, gill-over-the-ground, field balm, creeping charlie, cat's-foot, runaway-robin, ale-hoof, gillale

SOME FACTS Perennial; reproduces by rootstocks and by seeds; introduced from Eurasia

RANGE Widespread throughout northern United States and southern Canada; also in American South

HABITAT Damp, shaded ground

SEASON April–July

Prunella
Genus

Prunella vulgaris
▶ **Heal-all, selfheal, carpenter weed, prunella, sicklewort, heart-of-the-earth**

Heal-all is one of the very few weeds that almost every gardener in the world may find growing on his land, usually in poorly drained soils. Sicklewort is native to the United States, and to Europe and Asia as well, a really ubiquitous weed. If you happen to be in England, you might check on John Gerard's statement that *Prunella* "grows in Essex neere Heningham castle" and that those specimens were white-flowered.

This plant is not entirely without charm. Its blue-violet (sometimes white or pink) tubular sessile flowers are two-lipped, as are members of the mint family; the upper lip is arched and the lower lip is spreading and three-lobed. These blooms occur within a thick spike within the axils of bractlike leaves.

Depending on growth conditions, carpenter weed may be two inches to one foot high. In the sun the plant is short and darker green than in the shade. Constant mowing of a lawn infected with heal-all will cause horizontal spreading of the stems, which, since they can root at their nodes, will only spread the plant more rapidly.

Its leaves are oppositely disposed along the square stems so typical of mints, and are oblong-ovate, broader at their bases.

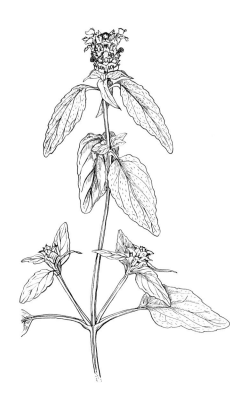

COMMON NAMES Heal-all, selfheal, carpenter weed, prunella, sicklewort, heart-of-the-earth

SOME FACTS Native of the United States and most of the rest of the world; perennial; reproduces by seeds

RANGE Widespread throughout North America

HABITAT Lawns, pastures, waste places, meadows

SEASON May–October

Eupatorium
Genus

Eupatorium purpureum
▶ **Joe-pye weed, purple boneset, tall boneset, trumpetweed, feverweed, queen-of-the-meadow, gravel-root, kidney-root**

There are a number of species of the genus *Eupatorium*, and one of them was used by Mithridates Eupator, King of Pontus, as an antidote for a certain poison. In finding the antidote he gave his name (which would otherwise have been forgotten) to the plant. *Eupatorium* is a composite and a close relative of the garden plant *Ageratum*.

The commonest common name, Joe-pye weed, comes directly from the name of an American Indian of our own Northeast. Joe-pye was an herbal doctor who specialized in tonics, decoctions, and medicines concocted from parts of this plant.

One of the taller weeds, this one may tower above you; its three-to-ten-feet smooth, ridged stems are capped with a great flat-topped corymb of purple heads. These stems are usually speckled with dark purple. Its large oblong-ovate to lanceolate leaves are arranged in whorls of three to six leaves each. The leaves are smooth above but hairs may be found along the veins of the underside.

Each purple head of the large flat corymb contains only tubular disc flowers. Unlike daisy or sunflower, no ray flowers are present in the head of purple boneset.

COMMON NAMES Joe-pye weed, purple boneset, tall boneset, trumpetweed, feverweed, queen-of-the-meadow, gravel-root, kidney-root

SOME FACTS Native to the United States; perennial; reproduces mainly by rhizomes, also by spores

RANGE There are two varieties, one common to the northeastern regions of the United States and one common to the Northwest; the latter is most troublesome west of the Cascades

HABITAT Chiefly on sandy or gravelly soils in upland pastures, abandoned fields

SEASON August–September

Asclepias
Genus

Asclepias syriaca
▶ **Milkweed, silkweed, cotton-
weed, silky swallowwort,
Virginia silk, wild cotton**

Asclepias was the god of medicine to the
Greeks (he was called Aesculapius by the
Romans). His name was given the plant
because it was once used in a number of
medicinal concoctions; it is still used in a for-
mula supposed to help the asthmatic.

Animals that ingest its leaves, and a
resinoid contained in them, show a number
of frightening symptoms—profound depres-
sion and weakness, spasms, labored respiration, elevation of the temperature with
dilation of the pupils (i.e., all the symptoms a
man in love endures), and then, finally, a com-
atose state ending in death.

Interestingly, despite these frightening
facts, the young shoots make a delightful
asparagus substitute if cooked. Boiling in
water removes the dangerous principle. Even
the older leaves may be cooked (after being
rinsed a number of times like spinach), and
they taste very good. The American Indians
were fond of cooked unopened flower buds
of this plant, and the very young pods are
cooked and eaten very much the way okra is.

COMMON NAMES Milkweed, silkweed,
cottonweed, silky swallowwort, Virginia
silk, wild cotton

SOME FACTS Native; perennial;
reproduces by seeds and by rhizome

RANGE Common in eastern North
America, west to Iowa and Kansas

HABITAT Fields, pastures, waste places,
roadsides; often on rich, gravelly or sandy
loam

SEASON June–August

Family

Arctium
Genus

Arctium lappa
▶ **Great burdock, beggar's buttons, cockle-button, clotbur**

The strikingly large leaves of the first year's rosette may grow to over one foot long; their lower surfaces are light green and woolly, while their upper surfaces are darker and smooth. In this species the petioles are solid. The first year's leaves arise from a short stem attached to a large, robust, deeply buried taproot that may be a foot long and three inches wide.

In the second growing season of its two-year life clotbur sends up a tall, flowering stem that is thick, hollow, and grooved. It branches and produces alternate leaves, which are smaller in size than the leaves of the earlier rosette, but of the same general shape. In the axils of the upper leaves on this rough-hairy stem are borne the clusters of floral heads. *Arctium* is also a composite.

Each nearly globular head may be more than an inch across and is usually on a longish peduncle. Individual flowers are pink-purple and tubular—there are no ray flowers—and have all their floral organs, including purple anthers and white stigmas on the pistil. Each bract of the involucre is hooked.

The burs attach themselves with great ease to animal fur as well as to the clothes of passing human beings. Elementary school boys are often guilty of throwing the burs into elementary school girls' long tresses, or were in the days when life was still simple and sweet at that age.

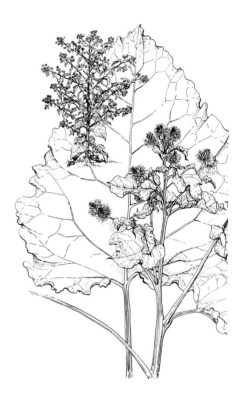

COMMON NAMES Great burdock, beggar's buttons, cockle-button, clotbur

SOME FACTS Introduced from Europe; biennial; reproduces by seeds

RANGE Widespread in the northeastern and north central states, and here and there on the Pacific Coast

HABITAT Roadsides, waste places, fence rows, neglected farmyards

SEASON July–October

Desmodium
Genus

Desmodium canadense
▶ **Tick trefoil, showy tick trefoil, stick-tight**

There are many species of the genus *Desmodium* but all are recognizable by their pods. The pod of *D. canadense*, showy tick trefoil, is about an inch long and is slightly curved. It contains three to five seeds and is covered with tiny hooked hairs that cause it to stick tightly to your clothing if you brush against the mature pods during one of your nature walks. The pods usually break into sections, each one a small oval or triangle containing one seed.

Showy tick trefoil produces typically leguminaceous flowers in racemes at the top of the branched stem. The racemes are densely flowered and contain ovate-lanceolate bracts. Most blossoms are bluish-purple, though some are nearly white. Bluish-purple flowers are rose-purple before opening, gradually assuming a darker hue.

The plant's erect, stout stem is branched above, and is ridged and grooved and very hairy. Each leaflet of the trifoliate leaves that are borne on the stem is oblong-ovate with a nearly smooth upper surface, though the lower surfaces are covered with finely appressed hairs. Each leaflet has numerous nearly straight veins. Leaves produced on the upper branches are nearly sessile.

COMMON NAMES Tick trefoil, showy tick trefoil, stick-tight

SOME FACTS Native to the United States; perennial; reproduces by seeds

RANGE Generally in eastern North America, south to the Carolinas

HABITAT Borders of woodlands and streams, old fields; often on gravelly soils

SEASON August–September

Family

Lythrum
Genus

Lythrum salicaria
▶ **Purple loosestrife, willow herb, spiked loosestrife, bouquet-violet**

Some readers may question the inclusion of purple loosestrife among the weedy species of the flowering plants—especially someone who has done what I have done: pulled off the road just to stop and stare at the flaming beauty of a pond in early autumn when it seems one mass of purple. Yet, this spectacle suggests a reason for inclusion. The very abundance in which the plant grows prevents other plants from having *Lebensraum*. Purple loosestrife literally chokes out native vegetation. And that is weediness. The definition of weed is not entirely based on disruption of the farm, garden, or lawn.

Purple loosestrife usually stands with its feet in water or in very wet soil. The tall stems of willow herb may sometimes reach four feet, and along this stem are borne opposite, sessile, lanceolate to nearly oblong leaves that taper to a point and may have white hairs on them. Sometimes the leaves are whorled, usually with three leaves at a node.

Its stunning blooms are produced within the axils of the leaves, leaves that are willow-like and give the plant its Latin specific epithet, *salicaria*, and its common name, willow herb. Its generic name refers to the bloodlike color of its flowers: *Lythrum* is derived from the Greek *lythron*, meaning "blood."

COMMON NAMES Purple loosestrife, willow herb, spiked loosestrife, bouquet-violet

SOME FACTS Perennial; reproduces by seeds; arrived from Europe

RANGE Northeastern United States to Virginia and Missouri; also on Pacific Coast

HABITAT Marshes, along shores and ponds, wet soil

SEASON June–September

Equisetum
Genus

Equisetum arvense
▶ **Field horsetail, scouring-rush, pinegrass, foxtail rush**

The field horsetail is common enough where it is not wanted to be a weed. Scouring rush produces a cone atop a jointed, yellowish, fluted stem; in the cone are formed thousands of small green four-winged spores. The cone looks like a papal tiara in overall shape.

After the spores have been shed, the fertile yellowing cone-bearing shoots die back to the jointed, perennial rhizomes, and then sterile green shoots come up from the rhizome. These shoots are eight inches to one foot in height, and are jointed and fluted, as were the fertile shoots; they are hollow between nodes and solid at the nodes, where sheaths composed of scalelike leaves that are fused at their bases are formed. The sheaths' tips appear as small "teeth" around the node.

What may unknowingly be interpreted as long, thin green leaves are actually branches, and they occur in whorls at the solid nodes of the main stem; each thin stem may also give rise to still thinner stems at each node, and thus a plumose plant that looks something like a horse's tail will be seen. Sterile shoots remain green and alive until autumn.

Horsetail contains one or more thiamine-destroying principles. Horses are particularly susceptible to poisoning by and have died of eating this plant. Administration of thiamine brings the animal back to health, unless it has already reached the last stages of this disease, known as equisetosis

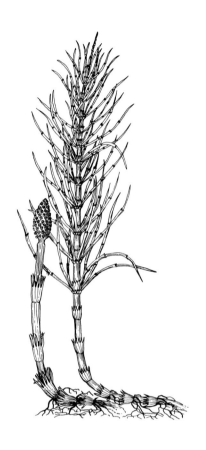

COMMON NAMES Field horsetail, scouring-rush, horsetail fern, pinegrass, foxtail rush, bottle brush, horsepipes, pinetop

SOME FACTS Native to United States and southern Canada; perennial; reproduces by spores (not seeds), creeping rootstocks, and tubers

RANGE Throughout the United States and southern Canada

HABITAT Moist fields, meadows, moist road embankments

SEASON April–May

Pteridium

Genus

Pteridium aquilinum
▶ **Eagle fern, bracken, brake fern, upland fern, hog brake**

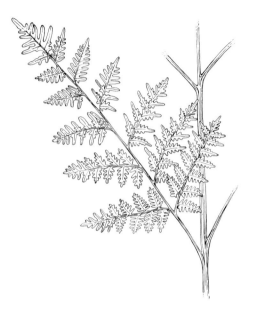

Pteridium aquilinum, the bracken fern, is the weediest of the ferns. Few ferns can tolerate bright sunlight. They are usually plants of shady nooks, but the eagle fern is quite tolerant of bright sunlight.

The colonies of many individuals, or what appear to be individuals, are not usually the products of a sexual reproductive process but have resulted from the continued growth of the rhizome, which may be some twenty feet long. Spores are produced along the margins of the frond, and before the sporangia are mature, the margin is bent under to cover and protect them during their development. When ready to shed their spore cargoes, the margins roll forward, giving the sporangia access to the open air.

Each frond arising from the rhizome is composed of three segments, each segment being twice pinnate. The frond may be one and one-half to four feet high.

Boiled in salted water, the unwound croziers, which are grayish and woolly, are edible. The rootstocks have been used as a food, too, and even to make a kind of rootbeer. Ingestion by stock of the uncooked, mature fronds is a serious matter, as they contain an enzyme that destroys thiamine (vitamin B), leading to serious illnesses in cattle, sheep, and other animals.

COMMON NAMES Eagle fern, bracken, brake fern, upland fern, hog brake, turkey-foot brake

SOME FACTS Native to the United States; perennial; reproduces mainly by rhizomes, also by spores

RANGE There are two varieties, one common to the northeastern regions of the United States and one common to the Northwest; the latter is most troublesome west of the Cascades

HABITAT Chiefly on sandy or gravelly soils in upland pastures, abandoned fields

Family

Glossary

A

achene A dry indehiscent (nonopening) one-seeded fruit common among the composites, though also found in other families; the sunflower "seed" is an achene, as is the true fruit of the strawberry

alkaloid Metabolic products of plants that are nitrogen-containing bases; some are poisonous; nicotine is an alkaloid

alternate Any arrangement of leaves or buds on a stem in which each is placed singly and at a different height

angiosperm The flowering plants; their seeds are "housed" in an ovary

annual Of one season's duration from seed to maturity and death

appressed Closely and flatly pressed against

ascending Stems curving upward from the base

awns A bristlelike part of appendage; found especially in the grass family

axil Upper angle where the leaf joins the stem

B

berry A fleshy fruit containing two or more seeds (usually more); e.g., tomato, grape, orange

biennial Of two season's duration from seed to maturity and death; flowering usually occurs during second year

bilaterally symmetrical Capable of division into two similar sections that are mirror images of each other; said of a flower; e.g., snapdragon, orchid

bipinnate Twice pinnately compound

blade The flat, expanded portion of a leaf (as contrasted with the petiole); the photosynthetic portion of the leaf

bract A much-reduced leaf, particularly the small scalelike leaves in a flower cluster; a small rudimentary or imperfectly developed leaf, usually green, though sometimes it is expanded and brightly colored

bulb A bud with fleshy bracts or scales, usually subterranean; common to many members of the lily family

C

calyx All the sepals of a flower collectively; the outer set of sterile floral leaves

cane The narrow flexible stem of small fruit-bearing plants; e.g., blackberry; the hollow jointed stem of taller grasses; e.g., giant reed grass

capsule A dry fruit of two or more carpels, usually dehiscent by valves splitting open along two or more divisions (sutures); e.g., the fruit of the poppy and evening primrose

caryopsis The fruit of a member of the grass family, a grain; containing one seed fused entirely to the ovary wall. The fusion is so perfect that this fruit (ripened ovary) is popularly called a seed. A seed is a ripened ovule. Every major civilization is based on the grains or caryopses of some member of the grass family

chlorophyll The green photosynthetic pigment found in the chloroplasts of plant cells; for photosynthesis to occur, chlorophyll must be present; thus this chemical is the source of all animal food and life

clasping Partly or wholly surrounding the stem, as among the monocots, where the leaf lacks a petiole, though many dicots may have leaves that clasp

composite Members of the Compositae whose inflorescence is a head or capitulum; e.g., dandelion, chicory, sunflower, or the garden aster, dahlia, daisy

compound leaf Leaves on which the blade consists of two or more separate parts called leaflets (*See also* Pinnately compound and Palmately compound)

corm An underground stem, bulblike in outward appearance, but with a solid interior; e.g., gladiolus

corn Once the general name for various grains (or grain plants), now, in the United States, usually limited to the corn plant (really maize plant), *Zea mays*, or to its ear of fruits; civilization in the Western Hemisphere has been based on this grain for thousands of years

corolla The inner set of sterile, usually colored floral leaves; the petals taken collectively

corymb A raceme with the lower flower stalks longer than those above so that all the flowers are seen on one level

cotyledon Seed leaf; the primary leaf (or leaves) of a plant embryo. In the monocotyledonae there is one seed leaf; in the dicotyledonae there are two

cruciform Having the shape of a cross, usually with four equal prongs

culm The stem of grasses, usually hollow except at the nodes, which are swollen

cyanogenic Giving rise to hydrogen cyanide (HCN), a deadly poisonous gas

cyme A broad, more or less flat-topped determinate flower cluster with the central flowers opening first; most often found among the members of the pink and borage families

D

deciduous Dying back; seasonal shedding of leaves or other structures; falling early

decumbent Lying flat or being prostrate, with the growing tip upward

dehiscence The splitting open of a fruit or anther to release seeds or pollen

deltoid Triangular or delta-shaped

dentate Toothed

dichotomous Forking regularly in (usually) equal pairs

digitate Diverging like the fingers of the hand

dioecious Having male and female flowers on totally different plants

disc (disk) flower The tubular flowers in the center of the head of many members of the composite family in contrast to the ray flowers around them; e.g., the yellow center of the daisy is composed of these flowers

dissected Divided into many segments

dorsal Back; back or outer surface of a part or organ

drupe A single-seeded fleshy fruit; e.g., cherry or peach

E

entire Used of a leaf that is neither divided nor toothed

epidermis The outermost tissue of leaves, young roots and stems, and other plant organs

evergreen Remaining green in all seasons. This term may be applied to flowering plants as well as to the conifers; many members of the heath family, such as rhododendron and mountain laurel, are evergreen

F

filament The stalk supporting the anther

floret A single small flower of a flower cluster such as is seen in the head of the composite, or a single flower of a grass

follicle A many-seeded dry fruit derived from a single carpel that splits longitudinally on but one side; e.g., milkweed

frond The leaf of a fern

G

glabrous Smooth or hairless

glaucus Covered with bluish or white bloom (wax)

globose Spherical or nearly spherical

glume One of the scaly bracts found at the base of and enclosing the spikelets of members of the grass family

grain The small, hard, one-seeded fruit of a grass; e.g., wheat, rice, corn, millet, barley grains

H

head A dense inflorescence of sessile or nearly sessile flowers on a very short axis, as seen among the composites; a capitulum; however, the flower of the clover, a legume, is also a head, though clover is not related to the composites

herb Lacking a woody nature; a plant naturally dying to the ground in winter

herbaceous Not woody; dying down each year

hirsute Having rather stiff, coarse hairs

I

imbricated Overlapping like the shingles of a roof

imperfect (flower) Either the stamens or the carpels are lacking

indehiscent Not regularly opening (as of seed pod or anther)

inflorescence The arrangement of several flowers on a flowering shoot (as in raceme, spike, head, etc.)

internode The part of the stem between two nodes (leaves)

L

lanceolate Several times longer than wide; broadest near the base, narrow at the apex

leaflet One part of a compound leaf

legume Dehiscent dry fruit of a simple pistil normally splitting along two sides

lemma The outer bract surrounding the grass flower

M

mechanical Said of plant tissue when it is supportive and strong

micron One one-thousandth (1/1000) of a millimeter

midrib The main vein of a leaf and most apparent (when apparent) in the leaf of a dicot

monocotyledonous A plant bearing one seed leaf (or monocotyledon); botanists shorten this to monocot; monocotyledonous plants have parallel veins within their sword-shaped, clasping leaves

monoecious Staminate and pistillate flowers on the same plant

N

net-veined Said of a leaf in which the principal veins form a network, as in the dicot leaf; such venation is also called reticulate

node A joint or place where a leaf or leaves are attached to a stem

O

opposite Two leaves or buds at a node; e.g., the maple or the ash

orbicular Circular

ovate With an outline like that of a hen's egg (broad end toward base)

ovule Undeveloped or immature unripened seed, found within ovaries of the angiosperms

P

palea The upper of two bracts enclosing the grass flower

palmate Radiating like a fan from approximately one point

palmately compound Said of a leaf when the leaflets radiate from one point (usually the distal end of the petiole); e.g., the leaf of Virginia creeper

panicle An inflorescence, a branched raceme with each branching bearing a raceme of flowers, all, usually, of a pyramidal shape

papilionaceous (corolla) Butterflylike in shape, as in the flower of the pea and bean; e.g., *Lotus*

pappus A ring of fine hairs developed from the sepals covering the fruits, especially among the composites; it is the pappus that acts as the "parachute" on the small achene of the dandelion

pedicel Stem of an individual flower in a cluster

peduncle Stem of a solitary flower or of a flower cluster

peltate Shield-shaped with its stalk attached within the margin

perennial Growing many years or seasons

perfect A flower with both stamens and pistils

perfoliate The stem appearing to pass through the leaf

perianth The calyx and corolla together

petal One of the modified leaves of the corolla (in animal-pollinated flowers usually colorful)

petiole The stalk of a leaf; in celery you eat the petiole

pinnately compound Leaflets arranged on each side of a common axis of a compound leaf; like a feather

pistillate Having pistils and no stamens; female

placenta The part or place in the ovary where the ovules find their attachment

plumule The bud or growing point of an embryo plant

pod A general term often used to designate a dry, opening fruit; e.g., a pea pod

pollen The microspores or grains (not in the sense of fruit—see Caryopsis) borne by the anthers; each grain contains the male elements or nuclei; pollen may be animal transported or carried by the water or wind

pome The fleshy fruit of apple, hawthorn, quince, etc.; a fruit with a bony or leathery several-celled core and a soft outer part

precumbent Trailing or lying flat on the ground

prickle A small sharp outgrowth of the epidermis; e.g., the rose "thorn" is really a prickle

privet Shrubby plants used as borders or as fencing; often members of the olive family, such as *Ligustrum*, though species of barberry (*Berberis*) may also be used; it means "private" and is meant to keep you that way

prostrate Lying flat on the ground

pteridophyte A fern or related plant

pubescent Covered with soft, fine hairs

R

raceme A simple flower cluster of pedicled units on a common elongated axis; the flowers open from the base upward

radial symmetry On looking down from above, the flower is circular and if any plane is passed through the center, mirror images (semicircles) are produced

ray flower A marginal flower with a strap-shaped corolla, as seen in the composites

receptacle The portion of the stem to which the floral organs are directly attached

reniform Kidney-shaped

reticulate The veins forming a network

rhizome An underground elongated stem

rosette A very short stem or axis bearing a dense cluster of leaves; e.g., the first year's growth of mullein

runner A slender trailing stem taking root at the nodes when it touches the ground

S

sagittate Arrowhead-shaped

samara An indehiscent winged fruit as seen on the tree-of-heaven (or tree-that-grows-in-Brooklyn) in the fall

sepal One member of the calyx

serrate Having sharp teeth pointing forward

sessile Lacking a petiole or stalk

spatulate Spoon-shaped; gradually narrowing downward from a rounded summit

spermatophyte Any seed-bearing plant

spike A flower cluster similar to a raceme with sessile or nearly sessile flowers

spikelet A small spike; the ultimate flower cluster in the inflorescence of a member of the grass family

spine A sharp, more or less woody outgrowth from a leaf or a completely modified leaf; e.g., the spines seen on a barberry (*See also* Thorn and Prickle)

stamen The pollen-bearing or male organ of a flower

staminate Having stamens and no pistils; male

stellate Star-shaped

stigma The part of the pistil that receives the pollen

stipule A basal appendage of a petiole; usually green

stolen A shoot that bends to the ground and takes root at the tip, giving rise to an entirely new plant

style The part of the pistil connecting the ovary and the stigma; usually, this part is more or less elongated

succulent Juicy; fleshy; soft and containing much water-storage tissue

T

taproot A root system with a main root bearing smaller lateral roots; e.g., the carrot

tendril A slender modified leaf or stem part by which a plant clings to a support

thorn A modified, degenerated sharp-pointed branch; true thorns are found on the hawthorn, but not on the rose (*See also* Prickle and Spine)

tomentose Covered with dense, woollike hair

trifoliate A compound leaf with three leaflets; e.g., clover

tuber Usually the enlarged end of a subterranean stem; the potato is a tuber

U

umbel An umbrellalike flower cluster found on members of the carrot family; e.g., Queen Anne's lace

V

ventral Front; relating to the inner face of an organ; opposite of dorsal

viscid Sticky

W

whorls Three or more leaves or buds at a node

wing A thin, dry membranaceous extension or flat extension of an organ; the later petals of a papilionaceous flower

Picture Credits

Larry Allain, USDA-NRCS PLANTS Database: Goosegrass

Jennifer Anderson, USDA-NRCS PLANTS Database: Cinquefoil, Poison Ivy, Bladder campion

N.L. Britton and A. Brown, *Illustrated Flora of the Northern States and Canada,* Courtesy of Kentucky Native Plant Society: Field bindweed, Fox grape, Butter and eggs, Common mullein, Scaldweed

Steven Daniel, Tick trefoil, Jewelweed, White campion, Blackberry, Dandelion, Wild cucumber, Vetch, Black-eyed Susan

William S. Justice, USDA-NRCS PLANTS Database: Yarrow, Swingle, Tree of heaven, Wild Garlic, Common Ragweed, Giant Ragweed, Chicory, Jimson weed, Wild carrot (Queen Anne), Cypress spurge, Ground ivy, Evening primrose, Narrow plantain, Black cherry, Black locust, Sheep sorrel, Curled dock, Nightshade, Chickweed, Tansy, Red clover, White clover, Moth mullein, Common mullein, Virginia Strawberry, Hawkweed

Dr. John Meade, Weed Scientist Emeritus, Rutgers Cooperative Extension: Japanese Knotweed, Yellow Foxtail, Purslane Flower, Pineapple Weed

Robert H. Mohlenbrock, USDA-NRCS PLANTS Database: Broomsedge bluestrem,

Yellow rocket, Devil's beggartick, Shepherd's purse, Canada thistle, Dayflower, Strawcolored flatsedge, Crabgrass, Barnyard grass, Horsetail, Boneset, Joe-Pye weed, Sunflower, Poverty rush, Bird's foot trefoil, Moneywort, Purple loostrife, White/yellow sweetclover, Carpetweed, Deer-tongue grass, Switchgrass, Timothy, Reed grass, Common Plantain, Knotweed, Poison ivy, Blackberry Bramble, Goldenrod, Goldenrod, Greenbrier, Greenbrier, Canadian goldenrod

John M. Randall, The Nature Conservancy: Downy bromegrass, Wild parsnip, Wild parsnip

James L. Reveal, USDA-NRCS PLANTS Database: Pokeweed, Cattails, Buttercup

George F. Russell, USDA-NRCS PLANTS Database: Ox-eye daisy, Japanese Honeysuckle, Milkweed, Brackenfern

Smithsonian Institute, Dept. of Systematic Biology, Botany, USDA-NRCS PLANTS Database: Wild morning glory

Carol A. Southby: Japanese Barberry, Viper's bugloss, Black-eyed Susan, Dandelions

Bill Summers, USDA-NRCS PLANTS Database: Lambs quarters

W.L. Wagner, USDA-NRCS PLANTS Database: Indian hemp, Burdock, Orchard grass, Selfheal

Index